Lesson Study Research and Practice in Mathematics Education

Lynn C. Hart • Alice S. Alston • Aki Murata
Editors

Lesson Study Research and Practice in Mathematics Education

Learning Together

Springer

Editors
Lynn C. Hart
Department of Early Childhood Education
Georgia State University
30303 Atlanta, Georgia
USA
lhart@gsu.edu

Aki Murata
School of Education
Stanford University
520 Galvez Mall
94305 Stanford, California
USA
akimura@stanford.edu

Alice S. Alston
Robert B. Davis Institute for Learning
Rutgers University
08901-1183 New Brunswick, New Jersey
USA
alston@rci.rutgers.edu

ISBN 978-90-481-9940-2 e-ISBN 978-90-481-9941-9
DOI 10.1007/978-90-481-9941-9
Springer Dordrecht Heidelberg London New York

Cover design: eStudio Calamar S.L.

Printed on acid-free paper

Springer is part of Springer Science+Business Media (www.springer.com).

Prologue

The title of this book intends to communicate two important beliefs. First, the book addresses *research and practice* in lesson study in 16 different locations. Although *research on lesson study* is still in its infancy, there are considerable efforts at *implementation of lesson study* occurring around the globe. In this book, we present these two critical areas, side-by-side, in an effort to gain a greater depth of understanding of the complexity of lesson study along with its benefits and challenges. We believe the two will inform each other.

Second, there is still much to be learned about lesson study outside Japan. For us, as editors of the book, the act of compiling the 16 stories presented in this volume provided us with a richer immersion into lesson study than could have been anticipated. By pulling together the various work that has been attempted and completed, and providing researchers and educators a tool for understanding lesson study, we are *learning together.*

The book expands the body of knowledge about lesson study and addresses questions that have been raised about lesson study. Lewis et al. (2009) state that "Efforts…to document the features and impact of lesson study have been modest to date" (p. 285). They go on to suggest that there are still questions that need to be addressed. They ask "…what are the features of lesson study; what are the mechanisms by which lesson study is posited to improve instruction; and what is the evidence that lesson study can be used effectively outside of Japan" (p. 286). The research and implementations described in this volume are the beginning of a response to those queries.

Identifying the features of lesson study is not a simple undertaking. As is seen in the chapters of this book, lesson study is a complex professional learning approach. We need to identify what is *essential* for an experience to be *lesson study*, rather than merely professional learning that is similar to lesson study. Parts IV and V in this volume explore that question.

In Part IV, Doig, Groves and Fujii use the metaphor of an iceberg to capture the unseen features of lesson study with respect to the task for exposing student thinking and impacting student learning. Their metaphor is useful (beyond the discussion of task) for lesson study. Just as the iceberg has much beneath the surface, many of the features (or essentials) of lesson study are not immediately obvious, and exposing them will assure fidelity of implementation of those essential features. Other essentials include the need for a learner stance (Sack and Vazquez) and teacher dispositions (Kamina and Tinto).

Authors in Part V add to the clarity of our vision of the iceberg. Sjostrom and Olson describe how they laid the groundwork for two years to support teacher dispositions for lesson study work. Bruce and Ladky provide rich images of the activities between the four stages that are necessary to accomplish the work of a lesson study. And Corcoran exposes how the power held by the outside expert in a traditional setting can impact the role of that person in a lesson study community.

Lewis' et al. (2009) second question "what are the mechanisms by which lesson study is posited to improve instruction" is addressed in the research studies in Part I. All of the studies in this part found that the practicing teachers engaged in lesson study developed new knowledge for teaching. Mechanisms that appear to support development of knowledge include discussing tasks (which exposes content and curricular knowledge through open discussion of (mis)understandings), exposing student thinking (so teachers can base future instructional decisions on that information), and reflecting on lessons (so the collective knowledge of the group could be brought to bear on revising the lesson).

Finally, Lewis et al. (2009) asked what evidence exists that lesson study can be used effectively outside of Japan. Although Part I is useful here also, Part II and III are particularly useful in thinking about this question. While none of the studies and implementations described in these parts was carried out for a sufficient time to see long-term impact, they all have encouraging results for the short term. Particularly encouraging is the use of lesson study in the preservice setting (Part II) and the university setting (Part III).

This book is another step in building the knowledge base on lesson study outside Japan. It does not answer all the questions. In fact, it raises more questions, but it is a stepping stone in building a solid foundation for our understanding of lesson study.

Lynn C. Hart
Alice S. Alston
Aki Murata

Contents

Contributors

Alice S. Alston Rutgers University, New Brunswick, NJ, USA
e-mail: alston@rci.rutgers.edu

Roya Basu Rahway Public Schools, Rahway, NJ, USA
e-mail: roya_enayati@yahoo.com

Megan Bomer Illinois Central College, East Peoria, IL, USA
e-mail: magoers@gmail.com

Catherine D. Bruce Trent University, Peterborough, Canada
e-mail: cathybruce@trentu.ca

Jane Carriere City Schools of Decatur, Decatur, GA, USA
e-mail: jane.carriere@comcast.com

Sandra Cooper Baylor University, Waco, TX, USA
e-mail: sandra_cooper@baylor.edu

Dolores Corcoran St Patrick's College, Dublin City University, Dublin, Ireland
e-mail: dolores.corcoran@spd.dcu.ie

Brian Doig Deakin University, Melbourne, Australia
e-mail: badoig@deakin.edu.au

Colleen M. Eddy University of North Texas, Denton, TX, USA
e-mail: Colleen.Eddy@unt.edu

Maria Lorelei Fernandez Florida International University, Miami, USA
e-mail: mfernan@fiu.edu

Toshiakira Fujii Tokyo Gakugei University, Tokyo, Japan
e-mail: tfujii@u-gakugei.ac.jp

Susie Groves Deakin University, Melbourne, Australia
e-mail: grovesac@deakin.edu.au

Lynn C. Hart Georgia State University, Atlanta, GA, USA
e-mail: lhart@gsu.edu

William C. Jackson Scarsdale Public Schools, Scarsdale, NJ, USA
e-mail: bjackson@scarsdaleschools.org

Debra L. Junk The University of Texas-Austin, Austin, TX, USA
e-mail: junkdeb@mail.utexas.edu

Michael Kamen Southwestern University, Georgetown, TX, USA
e-mail: kamenm@southwestern.edu

Penina Kamina SUNY College at Oneonta, Oneonta, NY, USA
e-mail: kaminapa@oneonta.edu

Andrea Knapp The University of Wisconsin-Stevens Point, Stevens Point, WI, USA
e-mail: andrea.knapp@uwsp.edu

Mary S. Ladky Trent University, Peterborough, Canada
e-mail: maryladky@trentu.ca

Catherine Lewis Mills College, Oakland, CA, USA
e-mail: clewis@mills.edu

Stephen Marble Southwestern University, Georgetown, TX, USA
e-mail: marble@southwestern.edu

Rachelle D. Meyer Baylor University, Waco, TX, USA
e-mail: rachelle_meyer@baylor.edu

Cynthia Moore Illinois State University, Normal, IL, USA
e-mail: cjmoor1@ilstu.edu

Kimberley P. Morris Plainfield Public Schools, South Plainfield, NJ, USA
e-mail: kmorris@plainfield.k12.nj.us

Joan Moss University of Toronto, Toronto, Canada
e-mail: jmoss@oise.utoronto.ca

Aki Murata Stanford University, Stanford, CA, USA
e-mail: akimura@stanford.edu

Jo Clay Olson Washington State University, Pullman, WA, USA
e-mail: jcolson@wsu.edu

Melfried Olson University of Hawaii at Mānoa, Honolulu, USA
e-mail: melfried@hawaii.edu

Lou Pedrick Rutgers University, New Brunswick, NJ, USA
e-mail: l_pedrick@yahoo.com

Despina Potari University of Athens, Athens, Greece
e-mail: dpotari@math.uoa.gr

Bindu E. Pothen Stanford University, Stanford, CA, USA
e-mail: bpothen@stanford.edu

Jacqueline Sack University of Houston Downtown, Houston, TX, USA
e-mail: sackj@uhd.edu

Cameron Sawyer Awatapu College, Palmerston North, New Zealand
e-mail: csawyer@rocketmail.com

Mary Pat Sjostrom Chaminade University of Honolulu, Honolulu, USA
e-mail: msjostro@chaminade.edu

Len Sparrow Curtin University, Perth, Australia
e-mail: l.sparrow@curtin.edu.au

Akihiko Takahashi DePaul University, Chicago, USA
e-mail: atakahas@depaul.edu

Diane Hobenshield Tepylo University of Toronto, Toronto, Canada
e-mail: diane.tepylo@utoronto.ca

Patricia Tinto Syracuse University, Syracuse, NY, USA
e-mail: pptinto@syr.edu

Irma Vazquez Houston Independent School District, Houston, TX, USA
e-mail: irmavazquez@sbcglobal.net

Tad Watanabe Kennesaw State University, Kennesaw, GA, USA
e-mail: twatanab@kennesaw.edu

Paul White Australian Catholic University, Sydney, Australia
e-mail: paul.white@acu.edu.au

Trena L. Wilkerson Baylor University, Waco, TX, USA
e-mail: trena_wilkerson@baylor.edu

Makoto Yoshida Center for Lesson Study, William Paterson University, Wayne, NJ, USA
e-mail: yoshidam@wpunj.edu

Paul Woo-Dong Yu Grand Valley State University, Allendale, MI, USA
e-mail: yupaul@gvsu.edu

Joseph Zilliox University of Hawaii, Honolulu, USA
e-mail: zilliox@hawaii.edu

Introduction: Conceptual Overview of Lesson Study

Aki Murata[*]

Lesson study is a collaboration-based teacher professional development approach that originated in Japan (Fernandez and Yoshida 2004; Lewis and Tsuchida 1998; Stigler and Hiebert 1999). Lesson study attracted the attention of an international audience in the past decade, and in 2002 it was one of the foci for the Ninth Conference of the International Congress on Mathematics Education (ICME). It subsequently spread to many other countries and more than a dozen international conferences and workshops were held around the world in which people shared their experiences and progress with lesson study as they adopted this new form of professional development in their unique cultural contexts (e.g., Conference on Learning Study 2006; Fujita et al. 2004; Lo 2003; National College for Educational Leadership 2004; Shimizu et al. 2005).

When first introduced in the United States in the late 1990s, lesson study quickly gathered U.S. educators' attention. It was considered an adoptable and effective innovation (Choksi and Fernandez 2004; Lesson Study Research Group 2007; Lewis et al. 2004; National Research Council 2002; North Regional Educational Laboratory 2002; Richardson 2004; Stepanek 2001, 2003; Takahashi and Yoshida 2004; Wilms 2003). A decade later, over 400 schools are involved in lesson study in the United States (Lesson Study Research Group 2007).

Despite the rapid rate of interest in this approach to professional development, lesson study remains relatively new to countries outside of Japan, and most schools and teachers are at the early stages of adoption and implementation of the innovation. And, while there is an emerging body of lesson study literature, we do not yet have a coherent and shared understanding of how lesson study effectively works in different contexts and models of teacher learning. The purpose of this chapter is to give a conceptual overview of lesson study, including its common structure (and variations), and its history. The chapter will also present emerging research

[*] Parts of this chapter appeared previously in International Encyclopedia of Education, Teacher Education volume.

A. Murata (✉)
Stanford University, Stanford, CA, USA
e-mail: akimura@stanford.edu

L. C. Hart et al. (eds.), *Lesson Study Research and Practice in Mathematics Education*,
DOI 10.1007/978-90-481-9941-9_1, © Springer Science+Business Media B.V. 2011

literature on the topic, address challenges in the field, as well as identify promising avenues for future research in lesson study.

Lesson Study: Structures, History, and Variation

Lesson study incorporates many characteristics of effective professional development programs identified in prior research: it is site-based, practice-oriented, focused on student learning, collaboration-based, and research-oriented (Bell and Gilbert 2004; Borko 2004; Cochran-Smith and Lytle 1999, 2001; Darling-Hammond 1994; Wang and O'Dell 2002; Little 2001; Hawley and Valli 1999; Wilson and Berne 1999). Lesson study places teachers at the center of the professional activity with their interests and a desire to better understand student learning based on their own teaching experiences. The idea is simple: teachers organically come together with a shared question regarding their students' learning, plan a lesson to make student learning visible, and examine and discuss what they observe. Through multiple iterations of the process, teachers have many opportunities to discuss student learning and how their teaching affects it. Lesson study typically follows the steps outlined in Fig. 1, with a research lesson (live lesson observation) as the centerpiece of the study process (Fernandez and Yoshida 2004; Lewis 2002; Lewis and Tsuchida 1998; Murata and Takahashi 2002; Wang-Iverson and Yoshida 2005).

After identifying a lesson goal, teachers plan a lesson. The goals can be general at first (e.g., how students understand equivalent fractions), and are increasingly refined and focused throughout the lesson study process to become specific research questions by the end (e.g., strategies students use to compare 2/4 and 3/6). Teachers

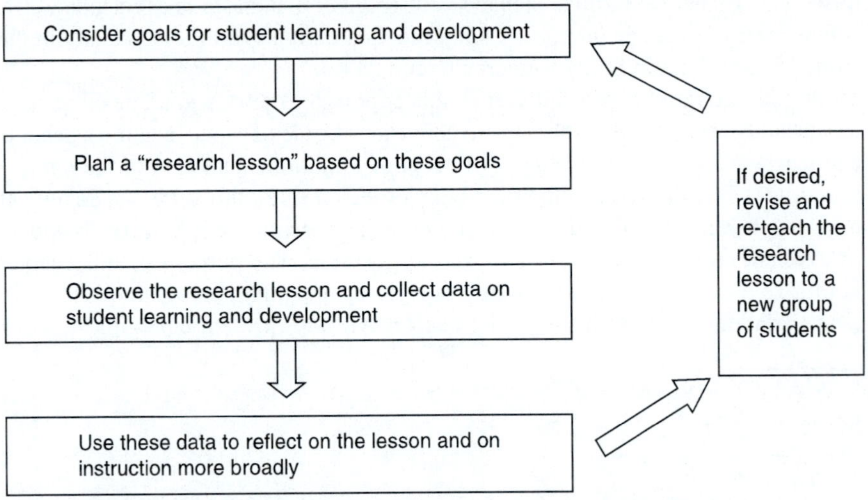

Fig. 1 Lesson study cycle

choose and/or design a teaching approach to make student learning visible, keeping their lesson goal in mind. The main purpose of this step is not to plan a perfect lesson but to test a teaching approach (or investigate a question about teaching) in a live context to study how students learn. As they plan, they anticipate students' possible responses and craft the details of the lesson. Teachers come to know the key aspects of the lesson, to anticipate how students may respond to these aspects, and to explore different thinking and reasoning that may lie behind the possible responses. During planning, teachers also have an opportunity to study curricular materials, which can help teachers' content knowledge development. During the lesson, teachers attend to student thinking and take notes on different student approaches. In the debriefing after the lesson, teachers discuss student learning based on the data they have collected during the observation.

There are other professional development programs that incorporate many of the characteristics of lesson study (e.g., action research, teacher research). However, what sets lesson study apart from those activities is the *live* research lesson. The live research lesson creates a unique learning opportunity for teachers. Shared classroom experiences expose teachers' professional knowledge that may otherwise not be shared: teachers notice certain aspects of teaching and learning. This implicit and organic noticing does not happen in artificially replicated professional development settings.

In Japan, lesson study has been widely used for over a century. Many Japanese educators attribute success in changing their teaching practice to participation in lesson study (Lewis et al. 2006; Murata and Takahashi 2002; Shimizu et al. 2005). As a foundational mechanism to support the improvement of teaching, lesson study is used to examine and better understand new educational approaches, curricular content, and instructional sequences introduced in Japan. In many cases, teachers play the central role in making new approaches adoptable and content accessible. Lesson study makes teaching approaches more practical and understandable to teachers through developing deeper understanding of content and student thinking. In this manner, lesson study works effectively to connect theory and practice in Japan.

While in the United States (and other parts of the world) lesson study is mainly known as a small, school-based collaboration, typically in the subject area of mathematics, it comes in many different shapes and sizes in Japan. There is small and school-based lesson study as well as large-scale, national-level lesson study (Murata and Takahashi 2002; Lewis and Tsuchida 1998; Shimizu et al. 2005). For a large-scale, national-level lesson study in Japan, teachers often travel long distances to participate, and hundreds of people can gather for one event. For mid-scale, district-level lesson study, teachers may come together for a district's professional development day where they have a menu of choices of lessons with different grade levels, subject areas[1], and topics to attend. The small-scale, in-school lesson study

[1] Lesson study research conducted outside of Japan has primarily been focused on mathematics so far, therefore the chapter reports findings from these studies on mathematics instruction. However, lesson study may be used for all subject areas.

(often emulated outside of Japan) is effective for teachers to improve their teaching for students within a particular community where teachers share knowledge of the students and the community.

Large-scale lesson study is important when a new educational approach (e.g., problem-based math instruction, collaborative learning), new content, or a sequence of content instruction is introduced and teachers across different schools try to make sense of what it means in their respective classrooms. Lesson study provides an opportunity to present an example of a new educational idea and/or approach for teachers to discuss, to ask questions about, and to construct a shared understanding of the new idea. Different forms of lesson study provide different learning opportunities for teachers. Different formats for lesson study meet different needs and interests of the teachers. A typical Japanese teacher has multiple opportunities to participate in lesson study throughout his/her professional career.

Focus on Student Learning

In the United States and other parts of the world, the new focus in mathematics teaching and learning requires teachers to balance and juggle existing knowledge of students, content, curriculum, and pedagogy while incorporating new ideas to make the practice conceptually stronger and more student centered (e.g., see National Council of Teachers of Mathematics 2000). Teaching is viewed as an interactive process in which student learning and content come together through effective teacher facilitation. This interactive teaching requires teachers to know how students *typically think* and *express their understanding* so that teachers can effectively facilitate their learning by weaving together different ideas. Teachers subsequently use this knowledge to provide experiences for students that encourage building connections among concepts and ideas. The focus on student learning binds different parts of the lesson study cycle, as teachers identify goals in terms of student learning of a topic, investigate curricular materials that teach the topic, plan a lesson to make student learning visible in the classroom with the topic, gather data in the lesson, and, afterwards, discuss the student learning that occurred during the lesson. Teachers become increasingly knowledgeable about a particular topic (content) and student learning of the topic in the process. They learn to listen to their students' ideas and to see student development.

One critical outcome of teacher learning in the process of lesson study is a new way to see teaching as a series of activities of inquiry around student learning. Lesson study helps cultivate a new attitude toward teaching, namely that teaching is not a one-way and didactic path, but a two-way integration of student ideas and content exploration meaningfully facilitated by teachers, an endeavor that can be extremely challenging. The emphasis on student learning in the lesson study process continually reminds teachers how important it is for them to understand students' ideas and helps bring the visions of reform into their classrooms.

Research Lesson: Centerpiece of Lesson Study

The research lesson is central to the lesson study process. In surveying 125 Japanese teachers, Murata and Takahashi (2002) found that teachers identified the research lesson as the most important element of lesson study that helped their professional growth. Through research lessons, teachers saw models of teaching and made sense of how the models affected student learning. The research lesson works to improve classroom practice, spread new content and approaches, connect classroom practice to broader educational goals, and explore conflicting ideas; thus creating demand for change, shaping national policy, and honoring the role of classroom teaching (Lewis and Tsuchida 1998).

Research lessons are observed in real time, and they provide a special learning opportunity within a developing professional community that teachers are not likely to find otherwise. Unlike watching a video segment of classroom teaching or reading teaching episodes in books, live lessons are experienced holistically. Events unfold as interactions among students and teacher(s) in the classroom develop. These events cannot be understood by independent analysis of each separate part (Davis and Summitt 2003; Herbst 2003). Classrooms are complex, and teachers/observers bring expert knowledge to understand this complexity. Through their unique lenses, they notice aspects of classroom experiences in their own ways and understand them as parts of a complex whole. They see relationships between small events that may not be visible to observers who have not spent time in classrooms as teachers. When experienced teachers come together and observe a live lesson, their expert knowledge comes to the surface as they interpret the effectiveness of the lesson and discuss it in the debriefing. The novice teachers who experience the lesson with experienced teachers are apprenticed into the profession through participation.

In some research lessons, knowledgeable others from outside the lesson study group are often invited to observe the lesson and provide comments and make suggestions for the research lesson at the end of the post-lesson discussion (Watanabe 2002). These commentators may be university researchers who specialize in the subject area addressed in the research lesson, or they may be experienced teachers who are interested in the topic. The main expectation for these commentators is to highlight special characteristics of the observed research lesson, tie them to research or theories of teaching-learning and/or conceptual development of students, and present a bigger picture of what their observations mean in the field of education. Unlike typical school consultants who observe and give feedback on aspects of teaching, focusing on what teachers should (or should not) do to make a particular lesson better, the research lesson commentator pulls together the different ideas and data shared in the debriefing to present a coherent picture of student learning. It requires good knowledge of the addressed topic, experiences in classrooms, and particular personal communication styles to be an effective outside commentator. With the short history of lesson study in countries other than Japan, this is an area that requires attention and development (Fernandez 2005; Lewis et al. 2006).

Research on Lesson Study

Teacher Learning

In examining the development and adaptation process of lesson study in the United States, Lewis et al. (2006) identified critical research needs, one being explication of the innovation mechanism. In order for us to understand how lesson study supports instructional improvement, we need to better understand what happens to teachers in its process. Initially in the United States, individuals interested in lesson study focused on the curricular resources (e.g., lesson plans) teachers produced as potential results of lesson study. While that was a reasonable expectation, after several years of lesson study effort, we are now in a better position to understand that in supporting instructional improvement, lesson study can produce much more than mere lesson plans. Murata et al. (2004) suggested three specific areas that develop and interact through the lesson study process to support teacher learning. The three broad areas, shown in the modified model in Fig. 2, are teachers' knowledge, teachers' commitment and community, and learning resources (see Lewis et al. 2006; Lewis et al. 2007).

Lesson study, according to Fernandez (2005), also provides opportunities for teachers to develop their pedagogical content knowledge. Different types of knowledge (e.g., knowledge of content, curricula, and student learning) come together and interact with one another during the lesson study cycle (Fig. 1). An ideal context is created in which teachers can integrate these types of knowledge and make content accessible to their students. Often in traditional professional development, these different types of knowledge are learned separately (e.g., attending a lecture on math content, reading a book on classroom management). In lesson study, they

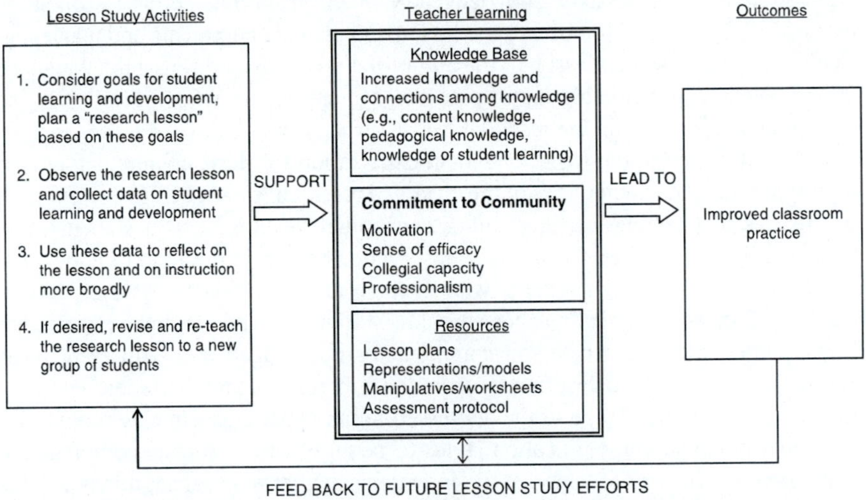

Fig. 2 Lesson study activities, teacher learning, and outcomes (modified by Murata et al. 2004)

come together and work interdependently to support student learning in the very practice of teaching, thus helping teachers experience different types of knowledge in a coherent and related whole.

The communities teachers create through lesson study support development of knowledge and connections among types of knowledge. While teaching is considered an independent and often isolated practice in many countries, lesson study brings teachers together to share goals, discuss ideas, and work collaboratively. It is likely that teachers who teach similar content to similar students have similar questions and issues about teaching. When these teachers gather and share their ideas and resources, a meaningful learning community is created, and the sense of belonging and professionalism developed in the community can strengthen teachers' commitment to their profession and motivate them to continually improve their practice (Grossman et al. 2001). For the teachers who collaboratively plan a research lesson, the process helps to add purpose to their everyday work. Their everyday life is experienced as a part of the larger professional endeavor among colleagues and an activity for which different events have clear purposes.

Obviously, the development and improvement of learning resources is a part of the lesson study process. Teachers' knowledge development and commitment to professional community growth interact with the development of learning resources (e.g., lesson plans), and as the resources are refined and improved, they provide a meaningful context for teachers to discuss student learning and to focus on the lesson. Just as young students find it helpful to have hands-on manipulatives to develop mathematical concepts, lesson plans become concrete scaffolds for teachers to focus their attention and learn about the specific content area under discussion.

The three areas supported through lesson study for teacher learning (knowledge development, community development, and material development; Fig. 2) are essential for instructional improvement and for increasing student achievement. Many professional development programs aim only to help teachers develop knowledge for teaching. While the single-focus approach is effective in some cases, when considering the sustainability of professional growth and teacher motivation, the three-part teacher-learning model identifies and incorporates the interactive relationships among different areas of teacher professional development.

Adapting Lesson Study

The Columbia University Teachers College Lesson Study Group presented several U.S. lesson study cases and identified central characteristics of participation that limited teachers' learning. When working with Japanese colleagues, U.S. teachers were challenged to find a strong research focus and to stay with the research process needed for lesson study (Fernandez et al. 2003). The U.S. teachers struggled to develop a meaningful research hypothesis, to develop means to explore the hypothesis, to use evidence to make claims, and to generalize the findings. In another study for which U.S. and Japanese teachers were interviewed about their lesson study experiences, the same group of researchers found that U.S. teachers were more likely to describe content goals (e.g., learning how to add fractions) in disconnect to other goals (e.g., student disposition) and focused heavily on what teachers do in lessons and not on student discovery and autonomy (Fernandez and Cannon 2005). Fernandez also investigated how teachers took advantage of learning opportunities that were created by lesson study (Fernandez 2005), and in the study, the lack of strong mathematics content knowledge and reasoning skills kept the U.S. teachers from taking full advantage of opportunities to learn. However, the author describes positive outcomes of how the teachers in the study collaboratively anticipated and discussed their students' thinking, revised and taught a lesson multiple times, and reflected on particular aspects of student thinking of mathematics that supported their learning as teachers.

One of the strengths of lesson study is that it places teachers' interests in the center of their learning process. In order for teachers to take full advantage of the opportunities of lesson study, they must be research-oriented and have inquisitive dispositions. However, if the teachers do not have these dispositions (as some research indicates), the dispositions can gradually be developed through participating in the lesson study process. Opportunities provided through lesson study support teachers as they develop knowledge and research skills and engage in future lesson study cycles in more effective and meaningful ways. While it may take longer for beginning lesson study participants to learn to hone in on the critical research process, in most cases, these teachers will become familiar and more adept with these expectations by their second or third lesson study experience. In the meantime, the sense of community and new professionalism will sustain their motivation to participate. Thus, these challenges found in the case studies mentioned earlier may be necessary learning steps for teachers who are for the first time considering teaching as a research process.

Issues from the Field

While the research on lesson study is in its early stages, there are numerous issues that have emerged from its implementation in the field. Many of these issues will be discussed in subsequent chapters of this book; four are worth mentioning here: (1) cost of implementation, (2) sustainability, (3) insufficient teacher content knowledge, and (4) connection to student learning. These four issues surface repeatedly in the literature on lesson study as an outcome of implementing lesson study in new cultural contexts. Frequently we lack the support and mechanisms required to handle the complexity of lesson study. As we become more adept with lesson study, the process through which we learn to handle these challenges needs to be documented and reflected on. Such an analysis will help expose strengths and weaknesses of our educational systems.

First, lesson study, as with any professional development, costs money to the districts and schools. While teachers may manage to find common planning time outside of their instructional time within school hours (e.g., lunch hours), shared observation for research lessons requires release time for the teachers, usually by hiring substitute teachers. If lesson study groups invite outside lesson study supporters (for ongoing support throughout the cycle) and/or knowledgeable others (for debriefs), that also may require additional cost. While external small grants are available for teachers, teachers (or district/school staff) will need to find time to gather resources, synthesize ideas, and write a proposal for the grant, which means additional work hours and cost.

Second, in most cultures outside of Japan (such as in the United States), professional development is usually experienced as discrete and separate programs from one year to another, but lesson study in Japan operates as a sustained effort that continues from year to year. In the new culture, teachers new to the lesson study process may consider enough to engage in lesson study for one year and then to move on to their next professional development experience. To be effective, lesson study should be rooted in a new culture as an integral part of teachers' professional lives. This may take some time and can be a challenge, as it may require the shift in cultural view toward professional development.

Third, lack of sufficient teacher content knowledge has been discussed in the previous studies (e.g., Ball 1990; Borko et al. 1992, Hart and Carriere this book), and we are also finding it difficult to support rich learning experiences when teachers who lack sufficient knowledge come to lesson study (Choksi and Fernandez 2004). As discussed in a section above, other studies found lesson study as a potential context to support knowledge development (e.g., Fernandez 2005), and some chapters of this book will also discuss findings in this area. We will need to find ways to make lesson study especially supportive of teacher knowledge development because the knowledge will provide the fuel for future learning with lesson study.

Finally, there is a strong need to make connections between teachers' lesson study participation and their students' learning. As the effectiveness of professional development is ultimately evaluated through its influence on student learning, we need a mechanism to show the connection between lesson study and student learning. In Japan, where lesson study has been used for many decades, they have enough experiences and data to show how instruction has improved and student learning shifted over time through lesson study, but in these new environments, careful documentation and various and creative investigations of student learning through lesson study will be necessary. It will not only help others understand the effectiveness of lesson study but also add a meaning to teachers' effort and professionalism.

Learning About Lesson Study Together

The chapters in this book will provide pictures of current efforts in research and educational activities with lesson study in different parts of the world. While we independently work and try to understand how lesson study works in our own context, it is likely that there are others who are having similar experiences. We need to find ways for us to communicate and share our emerging knowledge and understanding, so that we can better support our research and educational activities. This book is one such effort.

As the stories in this book unfold, we will examine how lesson study is used in different settings to support educational research and activities centered on teacher and student learning. We will have opportunities to learn what may or may not work well in new settings, and how different modifications and adjustments were made to better support the goals for each lesson study group. As teaching is a highly localized practice, modifications are expected and essential in order to adopt and use this new professional development approach effectively. Still, too many modifications may change the nature of lesson study, and teachers may find themselves participating in yet another ineffective professional development program with a new name. In order to avoid this, the key characteristics of lesson study should be maintained with care, while modifications are made. Some of these characteristics are listed below, however, I would caution the readers of this book that this is not an exhaustive list. In learning together about lesson study, I welcome others' ideas and input in expanding and modifying the list based on the localized knowledge gained though various contexts.

1. *Lesson study is centered around teachers' interests:* Teachers' interests are central to their professional development. Lesson study goals should be something teachers feel is important to investigate and relevant to their own classroom practice.
2. *Lesson study is student focused:* Lesson study is about student learning. At any part of the lesson study cycle (Fig. 1), the activities should focus teachers' attention to student learning and its connections to lessons/teaching.
3. *Lesson study has a research lesson:* Teachers have shared physical observation experiences (in some special cases, video may be used in place of the live lessons, but this is not recommended), that provide opportunities for teachers to be researchers.
4. *Lesson study is a reflective process:* Lesson study provides plenty of time and opportunities for teachers to reflect on their teaching practice and student learning, and the knowledge gained from and for the reflective practice should be shared in some format with the larger teaching and educational communities.
5. *Lesson study is collaborative:* Teachers work interdependently and collaboratively in lesson study.

By understanding how different aspects of lesson study may be modified while maintaining these key characteristics, we will better understand the existing educational system and cultural values and beliefs that support the system. When adding something new to an existing system, it often becomes clear what can and cannot be changed in the system to accommodate the new. That, in turn, helps us understand how different parts of the system work and what parts are more critical to the system than the others. Lesson study can provide that opportunity.

In the concluding chapter of this book, we will discuss possible next steps in our work with lesson study. We will reflect back on our work so far, identify common experiences and challenges, and present possible future paths and directions. What might have appeared to be discrete and unrelated experiences will hopefully begin to form a coherent picture with common purposes. The new understandings shared in this book will help drive our future work in this area with the goal of continuing to learn about lesson study together.

References

Ball, D. (1990). The mathematical understandings that prospective teachers bring to teacher education. *Elementary School Journal, 90*(4), 449–466.

Bell, B., & Gilbert, J. (2004). A model for achieving teacher development. In J. Gilbert (Ed.), *The Routledge Falmer reader in science education* (pp. 258–278). New York: Routledge Falmer.

Borko, H. (2004). Professional development and teacher learning: Mapping the terrain. *Educational Researcher, 33*(8), 3–15.

Borko, H., Eisenhart, M., Brown, C. A., Underhill, R. G., Jones, D., & Agard, P. C. (1992). Learning to teach hard mathematics: Do novice teachers and their instructors give up too easily? *Journal for Research in Mathematics Education, 23*(3), 194–222.

Cochran-Smith, M., & Lytle, S. (1999). The teacher research movement: A decade later. *Educational Researcher, 28*(7), 15–25.

Cochran-Smith, M., & Lytle, S. (2001). Beyond certainty: Taking an inquiry stance on practice. In A. Lieberman & L. Miller (Eds.), *Teachers caught in the action: Professional development in practice* (pp. 45–60). New York: Teachers College Press.

Conference on Learning Study. (2006). *Programme.* http://www.ied.edu.hk/cdspfe/lsconference/programme.htm. Accessed 19 Feb 2007. (The Hong Kong Institute of Education website).

Choksi, S., & Fernandez, C. (2004). Challenges to improving Japanese lesson study: Concerns, misconceptions, and nuances. *Phi Delta Kappan, 85*(7), 520–525.

Darling-Hammond, L. (1994). Developing professional development schools: Early lessons, challenges, and promise. In L. Darling-Hammond (Ed.), *Professional development schools: Schools for developing a profession* (pp. 1–27). New York: Teachers College Press.

Davis, B., & Summitt, E. (2003). Understanding learning systems: Mathematics education and complexity science. *Journal for Research in Mathematics Education, 34*, 137–167.

Fernandez, C. (2005). Lesson study: A means for elementary teachers to develop the knowledge of mathematics needed for reform-minded teaching? *Mathematical thinking and learning, 7*(4), 265–289.

Fernandez, C., & Cannon, J. (2005). What Japanese and U.S. teachers think about when constructing mathematics lessons: A preliminary investigation. *The Elementary School Journal, 105*(5), 481–498.

Fernandez, C., & Yoshida, M. (2004). Lesson study: A case of a Japanese approach to improving instruction through school-based teacher development. Mahwah: Lawrence Erlbaum.

Fernandez, C., Cannon, J., & Choksi, S. (2003). A U.S.—Japan lesson study collaborative reveals critical lenses for examining practice. *Teaching and Teacher Education, 19*, 171–185.

Fujita, H., Hashimoto, Y., Hodgson, B. R., Lee, P. Y., Lerman, S., & Sawada, T. (2004). *Proceedings of the ninth International Congress on Mathematics Education.* Norwell: Kluwer.

Grossman, P., Wineburg, S., & Woolworth, S. (2001). Toward a theory of teacher community. *Teachers College Record, 103*(6), 942–1012.

Herbst, P. (2003). Using novel tasks in teaching mathematics: Three tensions affecting the work of the teacher. *American Educational Research Journal, 40*, 197–238.

Hawley, W. D., & Valli, L. (1999). The essentials of effective professional development: A new consensus. In L. Darling-Hammond & G. Sykes (Eds.), *Teaching as the learning profession: Handbook for policy and practice* (pp. 127–150). San Francisco: Jossey-Bass.

Lesson Study Research Group. (2007). *Timeline of U.S. lesson study.* http://www.tc.edu/lesson-study/lsgroups.html. Accessed 15 Feb 2007. (Teachers College, Columbia University web site).

Lewis, C. (2002). *Lesson study: A handbook of teacher-led instructional change*. Philadelphia: Research for Better Schools.

Lewis, C., & Tsuchida, I. (1998). A lesson is like a swiftly flowing river: Research lessons and the improvement of Japanese education. *American Educator*, (Winter), 14–17, 50–52.

Lewis, C., Perry, R., & Hurd, J. (2004). Deeper look at lesson study. *Educational Leadership, 61*(5), 18–23.

Lewis, C., Perry, R., & Murata, A. (2006). How should research contribute to instructional improvement: The case of lesson study. *Educational researcher, 35*(3), 3–14.

Lewis, C., Perry, R., & Hurd, J. (2007). *Lesson study: A theoretical model and a North American case.* Paper in preparation.

Little, J. W. (2001). Professional development in pursuit of school reform. In A. Lieberman & L. Miller (Eds.), *Teachers caught in the action: Professional development that matters* (pp. 28–44). New York: Teachers College Press.

Lo, M. L. (2003). Lesson study and its impact on teacher development. *ISSP Newsletter*, 4–5.

Murata, A. & Takahashi, A. (2002). *District-level lesson study: How Japanese teachers improve their teaching of elementary mathematics.* Paper presented at a research pre-session of the annual meeting of the National Council of Teachers of Mathematics, Las Vegas, NV, USA.

Murata, A., Lewis, C., & Perry, R. (2004). Teacher learning and lesson study: Developing efficacy through experiencing student learning. In D. McDougall. (Ed.), *Proceedings of the twenty-sixth annual meeting of North American chapter of the international group of the Psychology of Mathematics Education* (pp. 985–992). Columbus: ERIC Clearinghouse for Science, Mathematics, and Environmental Education.

National College for School Leadership. (2004). Research lesson study: Experimenting with collaborative learning. *Networked Learning Group Publication, 6*, 30–31.

National Council of Teachers of Mathematics. (2000). *Principles and standards of school mathematics*. Reston: Author.

National Research Council. (2002). Studying classroom teaching as a medium for professional development: Proceedings of a U.S.—Japan Workshop. Washington: National Academy Press.

North Regional Educational Laboratory. (2002). *Teacher to teacher: Reshaping instruction through lesson study*. Naperville: North Central Regional Educational Laboratory.

Richardson, J. (2004). Lesson study: Teachers learn how to improve instruction. In Tools for Schools (pp. 1–3). Oxford, OH: National Staff Development Council.

Shimizu, S., Isoda, M., Okubo, K., & Baba, T. (Eds). (2005). *Mathematics lesson study in Japan through diagrams*. Tokyo: Meiji Tosho.

Stepanek, J. (2001). A new view of professional development. *Northwest Teacher, 2*(2), 2–5.

Stepanek, J. (2003). A lesson study team steps into the spotlight. *Northwest Teacher, 4*(3), 9–11.

Stigler, J. W. & Hiebert, J. (1999). The teaching gap: Best ideas from the world's teachers for improving education in the classroom. New York: Summit Books.

Takahashi, A., & Yoshida, M. (2004, May). Ideas for establishing lesson study communities. *Teaching Children Mathematics*, 436–443.

Wang, J., & O'Dell, S. J. (2002). Mentored learning to teach according to standards-based reform: A critical review. *Review of Educational Research, 72*(3), 481–546.

Wang-Iverson, P., & Yoshida, M. (2005). *Building our understanding of lesson study.* Philadelphia: Research for Better Schools.

Watanabe, T. (2002). Learning from Japanese lesson study. *Educational Leadership, 59*(6), 36–39.

Wilms, W. W. (2003). Altering the structure and culture of American public schools. *Phi Delta Kappan, 85*(8), 606–613.

Wilson, S. M., & Berne, J. (1999). Teacher learning and the acquisition of professional knowledge: An examination of research on contemporary professional development. *Review of Research in Education, 24*, 173–209.

Part I
Jumping into Lesson Study: Inservice Mathematics Teacher Education

Lesson Study: The Impact on Teachers' Knowledge for Teaching Mathematics

Rachelle D. Meyer and Trena L. Wilkerson

Rationale for the Study

Educational standards suggest that students should engage in complex problems that give rise to comprehensive mathematical understanding (National Council of Teachers of Mathematics 2000). Consequently, many teachers will have to shift their pedagogy of memorization, repetition, and recitation of correct answers to developing their students' reasoning and communication skills by actively engaging their students (Smith 2001). In the USA, professional development is often used to help teachers with pedagogical decisions and strategies for effective instruction as well as helping teachers understand the mathematics they teach. The most commonly used forms of professional development include short sessions at meetings, one-to-two day school-based workshops on specific topics, or two-to-three-week grant-supported workshops in the summer. However, another form of professional development occurring in the USA, on a much smaller scale, is lesson study. This study examined whether participating in lesson study as a form of professional development provides opportunities for teachers to improve their knowledge of teaching mathematics.

Much of education reform literature suggests that knowledge for teaching mathematics is essential to learning how to teach subject matter in order for students to understand it (Ma 1999; Smith et al. 2005; Hill et al. 2005). Conventional wisdom asserts, "You cannot teach what you don't know" (National Research Council 2001, p. 373). According to the National Research Council, an academically rich environment begins with teachers who are knowledgeable in mathematics, knowledgeable of students, and knowledgeable of instructional strategies. Knowledge of subject matter with an understanding of instruction results in a highly effective teacher (Phillips 2003; Hill et al. 2005). Research indicates that many US teachers do not possess a deep understanding of mathematics (Mewborn 2003; Ma 1999; National Research Council 2001) and evidence suggests that teachers, particularly at the elementary and

R. D. Meyer (✉)
Baylor University, Waco, TX, USA
e-mail: rachelle_meyer@baylor.edu

L. C. Hart et al. (eds.), *Lesson Study Research and Practice in Mathematics Education*,
DOI 10.1007/978-90-481-9941-9_2,

middle school grades, often have limited content knowledge (Smith 2001). Therefore, this study examined whether or not participation in lesson study would provide opportunities for teachers to improve their knowledge of teaching mathematics.

Context of the Study

Lesson study as defined by Lewis (2002) is a teacher-led instructional improvement cycle in which teachers work collaboratively to: formulate goals for student learning, plan a lesson, teach and/or observe the lesson, reflect on the gathered evidence, revise the lesson for improvement, and reteach the revised lesson (see Chap. 1 Murata). Through the use of lesson study, teachers have a means for planning, observing, and conferring with others.

Twenty-four mathematics teachers, from seven middle schools in a large urban school district, participated in an initial six-hour summer professional development session about lesson study with the goal to plan and implement a lesson study cycle by the end of the calendar year. Participants began by identifying a research theme/mission statement based on the qualities they would like their students to have in five years from now. This is an important first step because the theme should influence the planning of the research lesson. For example, if teachers want their students to be skilled problem solvers and communicators, then the lesson plan should support a discovery, trial and error approach versus direct instruction to become problem solvers. In addition, the plan should allow students to collaborate with peers as a way to develop communication skills. Next, school groups analyzed district curriculum expectations, district test results, and state assessment results in order to identify areas of need in mathematics. Each school group shared results with all participants in order to examine commonalities and differences in students' mathematical understandings for each of the seven schools. Then, each school group selected a content topic to address using the districts' scope and sequence for the second nine-week grading period as a guide. Participants regrouped themselves based on the topics chosen to be addressed in the research lessons with some teachers working outside of their school groups. Teachers also considered their greatest interest to learn and focus on teaching a lesson in this area. Five groups were formed consisting of four to five middle school mathematics teachers with some being from the same school while other groups were a combination of schools. Table 1 describes the demographics for each group.

The day continued with teachers gathering and studying resources on their identified content focus through the use of the internet, district textbooks and curriculum documents, standards-based curriculum resources, and other supporting elements provided by the researchers. Teachers were asked to focus on supporting skills needed to understand the content focus and the most effective sequencing of content topics. In addition, teachers were to assess their own understanding of the topic through discussion with their groups as they gathered information.

Next, the participating mathematics teachers were introduced to the history, goals, and procedures of lesson study. Participants viewed and reflected on a video

Table 1 Description of lesson study groups

Lesson study groups	Number of teachers in each group	Schools represented in each group	Content topic taught	Grade level
Group 1	4	1	Measurement	8th grade
Group 2	4	2	Proportional reasoning	8th grade
Group 3	6	2	Introduce proportional reasoning and scale factors	6th grade
Group 4	5	1	Theoretical and experimental probability	7th grade
Group 5	5	1	Area of similar figures	7th grade

of teachers who had participated in lesson study in order to see all phases of the lesson study cycle. Finally, each group was assigned a facilitator from a higher education institution. The higher education faculty member, whose expertise was in mathematics education, served as the facilitator when the groups met to continue planning their research lesson, and was available to answer questions when they arose. The facilitator was not to interfere with the direction of the research lesson. Groups concluded the initial meeting day by selecting a date for a two hour planning session to work on their research lesson.

Within three weeks following the initial six-hour professional development session, each group met for a two-hour planning session with their assigned facilitator and began working on their research lesson. During the planning session, the teachers completed a plan for the research lesson that included the following elements: the aim of the lesson (research theme/goal and the objective), the learning process for the lesson, the evaluation for the lesson, and copies of lesson materials (Lewis 2002).

The lesson plan consisted of three columns. The first column addressed the sequence of the lesson, problems to be posed, questions to be asked, and activities to be addressed. The second column consisted of anticipated students' questions and responses during various parts of the lesson. The third column was for specific aspects the lesson study team wanted observers to notice during the research lesson. In addition, the five groups determined which group member would voluntarily teach the research lesson. At the conclusion of the session, group members who would observe during the research lesson determined their observation roles and reviewed the observation procedures.

Observation roles consisted of a single observer recording comments and actions from a specific group of students. It should be noted that most lesson study teams had students work in collaborative groups with each group being assigned an observer. Additional observation roles consisted of recording all comments and questions asked by the teacher volunteer to implement the research lesson and an observer to randomly float around the room to the various student groups. Observers were also reminded not to interfere with the research lesson by talking with students and to record as much detail as possible for their observation focus.

On the day in which the research lesson was to be conducted, the volunteer teacher taught the lesson while team members and invited guests, such as the principal, wrote field notes regarding lesson implementation and student understanding. Following the research lesson, a two-hour reflection session on the research lesson was held. During the reflection session, the teacher who taught the lesson spoke first, reflecting on the lesson implementation, noting what went well, and on any difficulties in the lesson before others shared their reflections. Next, members who assisted in planning the lesson shared their observation notes reflecting on the goals for the students and the design of the research lesson, comparing and contrasting what was planned and what was observed. The discussion focused on the specific notes collected by each observer. Observation notes consisted of students and teacher comments and questions during the lesson as well as observation notes on the use of manipulatives and students' work. The group facilitator then provided feedback and shared in the discussion.

All participating groups were able to implement the research lessons by mid-December. In addition, the groups were able to modify the research lessons based on feedback during the reflection sessions and reteach the lessons. However, the reteaching of the lessons occurred in each individual teacher's classroom without observers due to limited resources for substitute teachers to cover classes for a second time. In early February, the five lesson study groups returned for a two-hour debriefing on their experiences and began plans for a second cycle of lesson study.

Analyzing the Impact of Lesson Study on Teachers' Knowledge

The researchers gathered data from the following sources to examine the impact participating in lesson study had on teachers' knowledge for teaching mathematics. These included: (a) transcripts from audio-recorded planning sessions, (b) research lesson plans, (c) transcripts from research lesson observation recording sheets, (d) transcripts from audio- and video-recorded reflection sessions of the participating teachers, and (e) teacher participant questionnaires. The analysis occurred in four stages. The first stage consisted of transcribing and verifying all records collected during the planning and reflection sessions as well as the research lesson observations. In the second stage two researchers independently hand coded the transcripts based on broad categories aligned with the research question such as "instructional practices" and "content knowledge." Third, the researchers met to discuss and verify the coding as a means to assure accuracy of the findings. The use of a computer software program, *N6* (Richards 2002), was used to code the emerging categories that emanated from the research questions. As patterns emerged, axial coding was used to identify subcategories. The fourth and final stage consisted of the researchers using a constant comparative analysis involving the multiple lesson study groups. This analysis consisted of both within- and across-group comparisons. For the within-group analysis, each group was first treated as a comprehensive study in and of itself. Once the analysis of each group was completed, a cross-group analysis was completed in order to develop more sophisticated descriptions and more powerful explanations (Merriam 1998).

What the Analysis Revealed

Analysis showed that particular factors of the lesson study cycle provided greater opportunity for teachers to increase their knowledge for teaching mathematics. These factors will be referred to as "windows of opportunity." One factor that impacted the development of teaching mathematics was whether or not the teachers implemented an existing lesson with little or no changes compared to creating or making major modifications to a task. A second factor that provided an opportunity to observe evidence of knowledge in teaching mathematics was when a group was able to anticipate students' questions/responses in the lesson plan (see Figure 1). A third factor which provided an opportunity to develop knowledge was when teachers took the time to discuss the content and not just the implementation of the lesson.

In the following discussion we will show the impact participation in lesson study had on the participating teachers. This will be supported through an analysis of statements by the participants throughout the planning and reflection sessions, the research lesson observations, and the lesson plans.

During the lesson study cycle, three windows of opportunity emerged and showed the potential for teachers to increase their knowledge for teaching mathematics. The following discussion will focus around these windows and how each group of teachers reacted to the opportunities. One factor included in the windows of opportunity is the lesson plan or task to be implemented. A second factor is the discussions teachers' had while planning the lessons. The third factor is the teachers' levels of anticipating students' questions and responses. We will begin by discussing the first two factors simultaneously.

Lesson Plans and Discussions

Teachers in groups 2, 3, and 4 showed an increase in their knowledge for teaching mathematics with the lesson task and discussions being contributing factors. Participants in groups 2 and 3 took an existing lesson plan and made significant modifications to the tasks, whereas group 4 created a new lesson task to be implemented.

During the initial planning, the teachers from group 2 decided to use an existing lesson on proportions, but made major changes in order to make mathematical and real world connections in addition to promoting student interest. For instance, teachers chose real world objects such as a cardboard box and a television to compare, in order to determine whether or not they were proportional. Teachers also spent significant time discussing their understanding of ratios and proportions. For example, one teacher asked, "What is the difference between a fraction and a ratio, and does it matter?" Rich questions and discussions such as this supported the teachers' opportunity to grow in their mathematical knowledge. By modifying the lesson plan, teachers had valuable discussions that allowed the individual knowledge of content and students to be expanded by the collective knowledge of the group.

Lesson Plan

Names:
Middle School: Grade:
Topic of the Lesson:
The Unit:
Research Theme (or "Main Aim") of Lesson Study:

Purpose (Goals/Objective):

Plan for the Research Lesson

Lesson Sequence (Questions, problems, and activities posed by the teacher)	Anticipated Student Questions/Responses	Points for Observers to Notice

Summary:
In Class Evaluation/Assessment:
Materials:
Extensions:

Fig. 1 Lesson plan format based on Lewis 2002

Teachers in group 3 also invested a significant amount of time and effort in creating a lesson that would allow students to conceptually understand the meaning of proportional reasoning through a discovery approach. Teachers decided to present a vignette where students would have to determine the height of their assigned "monster" based on a footprint and the use of nontraditional measuring tools. Teachers were aware of the development of their own mathematical knowledge by planning a discovery lesson. For example, one teacher commented, "By developing a problem for the students to solve, we were able to discuss what information the students would need, which ultimately strengthened my own understanding."

Results also indicated an opportunity to develop teachers' knowledge through the discussion of topics taught at each grade level and the instructional strategies used to teach such skills. For example, proportions were introduced at the 6th grade, and the teacher taught the "butterfly" method. However, the 7th-grade teacher explained his issues and concerns about using such a method to teach proportions, which resulted in the 6th-grade teacher increasing her own understanding of proportions. In fact, during the reflection session, this 6th-grade teacher said, "I do admit that having 7th- and 8th-grade teachers in the planning did help me see the vertical planning of ratios and proportions through the middle school years, and what I needed to know about proportions and how to teach it."

When designing the research lesson, the teachers from group 4 decided to choose an example to illustrate probability and fairness. The teachers decided to pose the question: Is *Rock, Paper, Scissors* a fair game? The teachers knew that their 7th-grade students often played this game to settle arguments, and students would be interested to find out if their opinions were supported mathematically. The facilitator later reflected that during the planning session, "This group seemed to really focus on probability and within probability, the fairness of a game. What fairness means and how to determine if it is present." The facilitator went on to say that in his expert opinion, teachers' mathematical knowledge is deepened when they see new and deeper connections regarding a topic in relation to the rest of the content, and these teachers illustrated that by their focus on the idea of fairness and the creation of a new lesson to address this idea.

While participating in lesson-study-provided opportunities to observe teachers' knowledge in teaching mathematics, groups 1 and 5 did not allow for this opportunity to be developed. Three factors may have contributed to this: (1) the use of an existing lesson with no modifications, (2) more focus on the implementation of the lesson and not the concepts, and (3) limited anticipation of students' questions. We will discuss the first two factors simultaneously.

Teachers from group 1 decided to utilize an existing lesson with no modifications and their focus became implementation and not the content of the lesson. For example, the teachers wanted students to increase their estimating and measuring skills so they used an established lesson activity where students would work in groups to estimate one person's facial features and then illustrate that face using accurate measurements. During the planning session, the teachers focused on the needed materials, the roles of the students, and who would teach the lesson with limited discussion on the content. The assigned facilitator attempted to guide the group by asking questions such as: "What do you want your students to understand?" "What do you want your students to be able to do?" "How will you know they can measure accurately?" "How will you know if the students understood?" One teacher responded, "We will look at the papers. If the nose is 30 cm then they are wrong." Another teacher said, "Maybe off two-three centimeters." The facilitator responded, "With a nose, that's a lot." Instead of continuing with a discussion about measurement, the teacher returned to discussing the implementation of the lesson. Another example occurred during the reflection session when the facilitator again questioned how the teachers would know when students' understood the content and one teacher responded, "A competition could be formed of who could draw the best; make the better features and so forth; which drawing came out the best." The teacher's focus was not on mathematical content but rather on nonmathematical areas such as artistic concerns and the ability to draw.

Teachers in group 5 did not benefit from the opportunities provided by lesson study to develop mathematical knowledge as a result of three contributing factors. First, only one teacher and the assistant principal attended the initial six-hour professional development session on lesson study. In addition, the assistant principal ended up writing the research lesson plan with limited input from the teachers. The teachers spent the entire planning session deciding on the topic of similar figures and sharing how they had taught the concept in the past. Finally, the teacher who volunteered to teach the research lesson did not follow the lesson plan but instead taught the topic as she previously had with her own students. Therefore, this group did not necessarily follow the lesson study cycle and this called into question the fidelity of implementation of lesson study.

The third factor in the windows of opportunity is the level in which teachers took the time to discuss and anticipate students' questions/responses. Again, teachers from groups 2, 3, and 4 showed an increase in their knowledge for teaching mathematics by anticipating students' questions and responses regarding the lesson.

Level of Anticipated Students' Questions/Responses

Teachers in group 2 spent a significant amount of time discussing and predicting students' questions and responses. There was one predicted question and seven anticipated responses. The predicted question pertained to mechanics such as, "What if only a little of a paper clip is left?" However, the anticipated student responses focused on conceptual understanding. For example, "Students should respond by saying that the taller the person, the larger his/her foot should be." Another prediction was, "Students may not understand which two columns [to use] from the chart to graph. They also may count boxes on the graph instead of lines and try to make a bar graph." The teachers also predicted, "Students will notice that the graph is [a] straight line and hopefully understand, if it is linear then it is proportional." Again, opportunities to observe teachers' knowledge of students and mathematics was provided through the discussion.

The teachers from group 3 put considerable thought into predicting students' questions. It should be noted that the teachers accurately predicted the student questions. For example, the teachers predicted numerous student questions regarding how to measure objects if not given standard measuring tools such as a ruler. "How do we figure this out?" "How am I supposed to figure this out without a ruler?" Analysis of the reflection transcripts from the research lesson reflected that students did indeed ask questions such as "How do we know if we are right without a ruler?"

Again, by having teachers predict students' questions and responses, this element encouraged teachers to think in terms of the students, which supports a better understanding of their own knowledge of both mathematics and students. One teacher even commented, "Trying to think like the students would think during the lesson made me have to look at the lesson as if I was learning it for the first time and even improved my own understanding."

On the lesson plan, teachers from group 4 also spent a significant amount of time discussing areas students may struggle with and were able to predict students' questions and responses that arose during the lesson task. For example, the teachers predicted that students would define fair as a 50/50 chance and would struggle with explaining fair if more than two people were involved. In addition, teach-

Table 2 Key factors of lesson study groups

Lesson study groups	Lesson plan/task	Level of anticipated students' questions/responses	Lesson-planning discussions
Group 1	Teachers used an existing lesson plan with no modifications from the original activity	Teachers listed five anticipated student questions that mainly addressed how to read a ruler. No student responses listed	Teachers focused on the implementation of the lesson and not on the content of the lesson
Group 2	Teachers used an existing lesson plan with major modifications connected to real world applications	Teachers listed one anticipated student question that pertained to instructions. Teachers listed seven responses that addressed students' level of conceptual understanding	Teachers focused on the implementation of the lesson and on the content of the lesson
Group 3	Teachers used an existing lesson plan with major modifications based on a discovery lesson	Teachers listed eight anticipated student questions that addressed areas students would struggle with based on a discovery approach	Teachers focused on the implementation of the lesson and on the content of the lesson
Group 4	Teachers created the lesson plan	Teachers listed two anticipated student questions that addressed rules for the game. Teachers listed five student responses that addressed students' level of conceptual understanding	Teachers focused on the implementation of the lesson and on the content of the lesson
Group 5	Lesson plan was created by the assistant principal with little teacher input	Teachers listed no anticipated student questions. Teachers listed three student responses that consisted of predicted students' answers to questions	Teachers focused on the implementation of the lesson and not on the content of the lesson

ers felt students could easily determine through tree diagrams whether or not the *Rock, Paper, Scissors* game was fair, but would struggle if the scenario changed. Instead of two people using either the rock, paper, or scissors symbol, would one player have an advantage if he/she never used the scissor symbol? Teachers were able to again anticipate what their students would do with the problem and their questions.

Unfortunately, teachers from groups 1 and 5 did not spend significant time discussing and predicting students' questions and responses to the lesson plan, and missed the third factor in the windows of opportunity to improve their knowledge for teaching mathematics. The teachers in group 5 did not discuss or write any student predictions and the teachers in group 1 only predicted student questions that addressed how to read a ruler. For example, they predicted, "What if the ruler is between two lines? How big is a centimeter?" The teachers did not discuss student responses or make predictions regarding their thoughts and their mistakes pertaining to conceptual understanding. Table 2 provides an overall summary of the lesson study groups and which factors did or did not improve teachers' knowledge in teaching mathematics.

Implications and Conclusion

Does participation in lesson study provide opportunities to improve the knowledge for teaching mathematics? Analysis suggests out of five lesson study groups, lesson study did provide opportunities for improved teacher knowledge in three of the groups. These groups displayed some common elements during their participation in lesson study that provided these opportunities.

First, all three groups planned research lessons that were either created or based on a previous lesson, but made significant changes in the lesson in order to address students' interests and make connections. On the other hand, the two groups who implemented previously established lessons with little or no modifications did not demonstrate these behaviors. We are not suggesting that teachers' knowledge cannot improve when using an established lesson plan for lesson study, but pointing out that this seemed to be a contributing factor for these groups.

Another common characteristic of the three groups was relatively accurate predictions regarding students' questions and responses. These groups had in-depth comments and questions listed under this particular column on the lesson plan that revealed careful thought. For example, one group predicted students would be able to state, "A fair game is when each person has an equal chance of winning and not just 50/50 always." The other lesson plans anticipated very few student questions and responses and appeared to be limited in their own depth of mathematical understanding and knowledge of students. For example, when estimating and measuring facial features, the teachers listed questions such as "What happens if it's between the lines?"

The third characteristic present in these three groups was the focus on the understanding of the mathematical concept and not only the implementation of the lesson.

Time was spent in their planning sessions talking about and discussing possible activities to help students conceptually understand and make mathematical connections regarding the research lesson focus. Whereas the other two groups focused on the procedural aspects of the lesson such as the location for the lesson to be taught, the grouping of the students, and who was going to do what to prepare for the lesson. While these may be valid areas to address when planning a lesson, they do not necessarily reflect a deep understanding of the knowledge needed for teaching mathematics.

This study was to explore whether participation in lesson study would (1) result in an increase in teachers' mathematical knowledge and (2) improve teachers' knowledge for teaching mathematics. More research is needed to examine the impact lesson study has on teachers' knowledge. However, this study provides insight into potential factors that would help support an increase in teachers' knowledge of teaching mathematics while participating in lesson study.

This study supports recent research by Ball et al. (2008) related to potential elements that increase teachers' knowledge of teaching. Analysis showed that teachers who participated in lesson study had opportunities to develop knowledge by predicting students' responses and questions. In addition, the teachers who discussed the targeted content areas and not just the implementation of the lesson revealed a deeper understanding for teaching mathematics. So does teachers' knowledge of teaching mathematics increase when they focus on anticipating students' questions and responses? Does teachers' mathematical knowledge increase with less focus on the implementation of the lesson and more on the activities and connections in mathematics and the real world? In three out of five lesson study groups, the answer seems to be that lesson study did provide "windows of opportunities" to observe and develop mathematical knowledge for teaching.

It is imperative for teachers to strive for continuous improvement of instructional strategies and knowledge because teachers are the key to students' understanding and achievement in mathematics (Dana and Yendol-Silva 2003) and instructional changes are more likely to occur in sustained efforts and in small incremental steps (Guskey 2000). Lesson study addresses one lesson at a time, but impacts learning and instruction in several aspects. Lesson study allows teachers to view teaching and learning as they occur in the classroom. With time, lesson study has the potential to build learning communities within schools and ultimately result in instructional improvement and increase in teachers' knowledge with focus on the student and the content.

References

Ball, D. L., Thames, M. H., & Phelps, G. (2008). Content knowledge for teaching: What makes it special? *Journal of Teacher Education, 59*(5), 389–407.

Dana, N. F., & Yendol-Silva, D. (2003). *The reflective educator's guide to classroom research: Learning to teach and teaching to learn through practitioner inquiry*. Thousand Oaks: Corwin Press.

Guskey, T. R. (2000). *Evaluating professional development*. Thousand Oaks: Corwin Press.

Hill, H. C., Rowan, B., & Ball, D. L. (2005). Effects of teachers' mathematical knowledge for teaching on student achievement. *American Education Research Journal, 42*(2), 371–406.

Lewis, C. (2002). *Lesson study: A handbook of teacher-led instructional change*. Philadelphia: Research for Better Schools.

Ma, L. (1999). *Knowing and teaching elementary mathematics*. Mahwah: Lawrence Erlbaum Associates.

Merriam, S. B. (1998). *Qualitative research and case study applications in education: Revised and expanded from case study research in education* (2nd ed.). San Francisco: Jossey-Bass.

Mewborn, D. S. (2003). Teaching, teachers' knowledge, and their professional development. In J. Kilpatrick, W. G. Martin, & D. Schifter (Eds.), *A research companion to principles and standards for school mathematics* (pp. 45–52). Reston: National Council of Teachers of Mathematics.

National Council of Teachers of Mathematics. (2000). *Principles and standards for school mathematics*. Reston: Author.

National Research Council. (2001). *Adding it up: Helping children learn mathematics*. (Edited by J. Kilpartrick, J. Swafford, & B. Findell). Washington: National Academy Press.

Phillips, J. (2003, Spring). Powerful learning: Creating learning communities in urban school reform. *Journal of Curriculum and Supervision, 18*(3), 240–258.

Richards, L. (2002, March). *Using N6 in qualitative research*. Melbourne: QSR International Pty Ltd.

Smith, M. S. (2001). *Practice-based professional development for teachers of mathematics*. Reston: The National Council of Teachers of Mathematics.

Smith, T. M., Desimone, L. M., & Ueno, K. (2005, Spring). Highly qualified to do what? The relationship between NCLB teacher quality mandates and the use of reform-oriented instruction in middle school mathematics. *Educational Evaluation and Policy Analysis, 27*(1), 75–109.

Developing the Habits of Mind for a Successful Lesson Study Community

Lynn C. Hart and Jane Carriere

Background

Significant differences in the curricula and approaches to professional learning of teachers in Japan and the United States, together with a lack of experience doing lesson study in the United States, make developing and maturing as a productive lesson study community a challenging process for U.S. teachers. For example, the entire national *Course of Study* for elementary schools in Japan is contained in a 100-page volume which lays out the hours, goals, and content for all 12 areas of study (including mathematics), allowing for invention and interpretation of best practice. In contrast, U.S. curricula cover many pages of objectives and skills for each content area at each grade level, allowing little interpretation for implementation by teachers. Professional learning activities that are designed to improve instruction are also in stark contrast in the two countries. Lesson study, the primary professional learning model in Japan is a teacher-driven and teacher-directed professional learning model. U.S. teachers experience most professional learning as top-down, outside-expert-directed (Lewis 2002). This chapter is divided into 2 parts. Part 1 is a detailed case description of a group of third-grade teachers in one school district in the United States as they embarked on the process of learning about lesson study and attempting to engage in the process. Part 2 is a research report on the group of teachers using data collected at the beginning and end of the yearlong process.

Theoretical Framework

Both the lesson study community and the research community benefited from the work of Fernandez et al. (2003), a study of implementation of lesson study in an urban, public school in New Jersey. They found "substantial challenges… must be

L. C. Hart (✉)
Georgia State University, Atlanta, GA, USA
e-mail: lhart@gsu.edu

L. C. Hart et al. (eds.), *Lesson Study Research and Practice in Mathematics Education*,
DOI 10.1007/978-90-481-9941-9_3,

overcome to make this practice [lesson study] purposeful and powerful" (p. 181). In their project, Japanese teachers from a nearby international school collaborated with U.S. teachers, attempting to form a lesson study community. The research team noticed that the Japanese teachers approached lesson study very differently than the U.S. teachers. In their analysis of the data they identified three habits of mind (which they named *Critical Lenses*) that guided how the Japanese teachers discussed and developed the lessons. They labeled these lenses as the researcher lens, which the teachers used to ask questions about practice and design classroom experiences to explore these questions; the student lens, which the teachers used to understand student thinking and examine all aspects of the lesson through the eyes of the student; and the curriculum developer lens, which the teachers used to organize, sequence, and connect learning experiences. They suggested that these critical lenses were absent in the U.S. teachers and prevented lesson study discussions from moving into rich arenas. Acquiring these lenses will be necessary if lesson study is to become a viable model of professional learning in the United States. In our project we use the three lenses identified by Fernandez et al. (2003) as a framework to both scaffold teacher thinking and ultimately study teacher change.

In addition, the role of beliefs in teacher change in mathematics education has a long history and is well documented (Cooney and Shealy 1997; Pajares 1992; Richardson 1996). Beliefs about *teaching* are well established by the time a student enters college. These beliefs, previously identified by Lortie (1975), develop during *the apprenticeship of observation* that occurs over years as a student. More importantly, we know that the beliefs teachers hold frequently do not support what Ma (1999) refers to as profound, *connected* understanding of fundamental mathematics teaching. In many instances, a teacher's beliefs determine the instructional decisions that are made in the classroom and limit willingness to engage in alternative pedagogies or behaviors. These beliefs are hard to change. Even in the face of conflicting evidence, beliefs are often resilient. A goal of professional learning, then, is not only to provide teachers with new frameworks to use as they plan learning experiences, but also to effectively challenge the core beliefs that drive their decision making.

Using these theoretical perspectives, we asked the following research question. Do teachers participating in their initial experiences with lesson study and supported by outside experts (facilitators) develop the lenses needed to be a productive lesson study community?

Overview

In this chapter, we provide a case of third-grade teachers as they develop a lesson study community (Part 1). The process was not easy for teachers raised in our Western educational system, and their struggles and accomplishments make the complexity of their task transparent. In particular, we detail the initial sessions to highlight the multifaceted learning process that was required. In the process, we describe our dual role as researchers and facilitators/participants within the group. In Part 2, we share results of analysis of the initial and final research lesson sessions to expose the development of the critical lenses.

Part 1: Becoming a Lesson Study Community

The School System

The school system in this project is a small urban system in the southern United States. During the period of the study, the system had six elementary schools. Thirty-eight percent of students were on free or reduced lunch. Fifty-three percent were African-American or other minority, and 47% were Caucasian. For many years, the system had been proactive in educating its teachers on reform mathematics as advocated by the National Council of Teachers of Mathematics (2000); however, as is the case in many school systems, attrition diminished the number of teachers who were prepared to teach from a reform perspective. New teachers, both new to the system and new to teaching, continued to bring a traditional way of thinking about the teaching of mathematics. According to the system mathematics coordinator, the curriculum and textbooks supported the philosophy of reform, but a teacher-directed model remained the primary mode of instruction in the classroom. The system sought to initiate a model of professional learning that could facilitate change and be self-sustaining.

Japanese Elementary Lesson Study (JELS)

The system chose to implement lesson study, starting with one grade level. The Japanese Elementary Lesson Study (JELS) Project was supported by a small external grant and funds from the school system. The project was facilitated by the mathematics coordinator for the school system and a local university mathematics education faculty member.

The Participants

All third-grade teachers from the school system were invited to participate in JELS. Participation was voluntary and eight of ten teachers opted to participate, representing five of the six elementary schools. There were two African-American females, one Asian male, and five Caucasian females. Teachers ranged from 3 to 30 years teaching experience.

The mathematics coordinator and university faculty consultant acted as facilitators/participants for the project. Both were white females and both had extensive practice with elementary teachers in developing standards-based learning environments. However, both were also new to the lesson study process and were new learners with the teachers.

Adapting Lesson Study

While preparing for the project, the project facilitators immediately noticed they would need to make adaptations to the lesson study process as it was described in the literature. There were several issues to be considered.

First, teachers would need to study and learn *about* the lesson study process. Unlike Japanese teachers who experience lesson study as a regular and on-going part of their professional development as early as their preservice program, for the U.S. teachers lesson study was new and not the type of professional learning they normally engaged in. The facilitators also held concerns about the readiness of the participants to engage in meaningful, substantive discussions about content and pedagogy with limited prior experience. To encourage profound, thoughtful discussion, the facilitators planned to be more active members of the lesson study group than described in the literature. They would scaffold the discussion through prompts such as "what if" and "did you notice" observations. This was supported by Fernandez et al. (2003) when they stated "lesson study…must include room for knowledgeable coaches who can stimulate the thinking of groups so they can rise beyond their own limitations" (p. 182).

Second, there would need to be adaptations to the Everyday Mathematics (EM) curriculum used in the system. The EM curriculum is relatively scripted and lessons are sequential and developmental. General objectives of a lesson could be used as a starting point, but it was anticipated that it would be necessary to revise the lesson to assure that student thinking was exposed and could be observed. We anticipated that it would be necessary for the group to develop single research lessons that could be inserted into the EM sequence. The teachers could identify concepts from EM they felt were difficult to teach and difficult to learn and develop introductory research lessons for those concepts.

Third, no common planning time was available that would allow the participating teachers to meet during the school day throughout the year. After school, conflicts prohibited using out-of-school time to meet. In-school time was necessary and substitutes were required. In order to schedule substitutes, full day sessions would be needed (a day to plan the lesson; a second day to teach/observe, debrief, and revise the lesson; and a third day to teach the revised lesson and reflect).

Finally, there existed a wide range of mathematical backgrounds of the participating teachers in the project. To engage in substantive discussions, substantive mathematics needed to be explored. The facilitators would develop a mathematics lesson for the *teachers to explore* the concept before they planned a research lesson for third graders. The exploratory lesson would immerse the teachers in the concept before they began the planning process.

Summer Sessions

The first of three summer professional learning days was held in June. On that day, teachers began to learn about lesson study. A video of a research lesson *Can you lift 100 kilograms?* (Lewis 2000) was used to introduce the process. After discussing the video, the group developed group norms for how they would work together. They developed a common vision of good teaching. They developed a list of long-term goals for students. Finally, they identified four troublesome topics specific to their grade level that they would address from EM: addition/subtraction word problems, the language of probability, comparing and ordering decimals, and naming parts of the whole in fractions. In preparation for the two professional learning days in August, the teachers were asked to read *Lesson study: A Handbook of Teacher-led Instructional Change* (Lewis 2002).

In August, the teachers met two days to discuss the book, develop a research theme, and plan the first of four research lessons for the year. This was daunting, slow work as the teachers were not accustomed to professional activities in which they were expected to not only share the authority, but to take the lead. They were not accustomed to considering or discussing what they thought were the important, over-arching learning goals for third-grade students.

Planning the Initial Research Lesson

Planning the research lesson consumed the remainder of the day and all of the next. The process began with one of the facilitators initiating a discussion on the mathematics that would be the center of the research lesson, i.e., thinking models for addition and subtraction word problems (join, separate, part/part/whole, and comparison), by using a "tell me what you know" approach. The teachers brainstormed their understandings of the eleven different forms the problems could take, discussing the relative difficulty of each type. After the discussion, the teachers examined existing EM curricula and materials, shared ideas from their own classrooms, looked at other resources, and collaboratively worked on developing a lesson. This too was difficult. The teacher participants had never formally shared pedagogical knowledge, negotiated teaching strategies, and discussed student learning and outcomes with colleagues. As teachers presented ideas that were not well received, group norms had to be revisited. How will we deal with differing opinions? How will we decide which strategy to use? Ultimately, the teachers agreed on a plan in which the students would solve one problem for each type of thinking model (join, separate, part/part/whole, and comparison). Students would discuss similarities and differences across the problems.

The group wrote the following four problems:

- There were 16 students in Mr. Bob's third-grade class. After winter break, he got three new students. How many students are in the class now? [*join*]
- Mr. Bob had 16 students in his third-grade class. Three students moved away. How many students are in the class now? [*separate*]

- Of the 16 students in Mr. Bob's class, seven are boys. How many girls are in the class? [*part/part/whole*]
- There are seven boys and nine girls in Mr. Bob's third-grade class. Of all the students in the class, how many more girls are there than boys? [*comparison*]

Teachers chose to use the same basic scenario and similar, simple numbers for the problems so the students would not get lost in the computation or the context. They wanted the students to focus on the differences in the thinking involved in the four problems types.

Students would be divided into small groups with chart paper divided into four sections. The four problems would be posted at different tables. Students would rotate from table to table with their chart paper and record their work in the appropriate section of the chart paper. Colored counting chips would be available at each table. When the rotations were complete the four charts would be posted side-by-side on the board and discussed, across all solutions for each problem number and then across the four problem types. The teacher was to focus the discussion on what was similar about the problems and what was different.

As recommended in the lesson study literature, the teachers agreed that observers would not interact with students. There was, however, discussion about whether observers would stay at a table or rotate with a student group. Pros and cons for both (understanding how all the children thought about one problem vs. how one set of children thought about all the problems) were discussed. In the end, the group opted to stay at one table in order to observe all students' thinking on one problem.

Once completed, a typed version of the lesson plan was created. The plan included teacher actions and strategies, anticipated student responses and thinking, and points for the observers to notice.

Organizing the Debriefing of the Research Lesson

The possibility of a public discussion about the lesson (the debriefing as a group) created some angst amongst the teachers. The facilitators reminded the teachers that the reflection was not a review of "how well the teacher had done" rather it was an assessment of how well the lesson exposed student thinking and how well it reached the objectives. It was agreed that once the group reconvened after the lesson, the teacher who taught would begin by sharing his or her observations and thoughts. The observing teachers would then share the data they had collected while observing the children. The facilitators would summarize the discussion, noting themes and issues that emerged, and the lesson would be revised based on observations shared during the debriefing. Planning for the first research lesson was completed on the last day of the summer meetings. Peg, a third-year teacher, volunteered to teach the first lesson. Subsequent lessons would be taught by the other teachers.

School-Year Sessions

The teaching of that initial lesson occurred three weeks after the school year began. Three additional cycles occurred over the year and followed the same process. The teachers would meet for one day to plan the lesson on the concept identified during the previous summer. On the next day, the lesson would be taught, debriefed, and revised. On the third day, the revised lesson would be taught to another group of third graders.

Teaching the Initial Lesson

The first 45-minute lesson was implemented during the third week of school. Peg, the teacher for this initial lesson, explained the purpose of the observers to the students, went over the problems with the students, put the children in groups, and set them to work. The observers watched the students and made notes.

Debriefing the Initial Lesson

After teaching the lesson, Peg began the debriefing. She expressed overall satisfaction with how the lesson proceeded, but acknowledged she was "more nervous than she expected to be" with observers in the room. She expressed concern that the lesson had not been as successful as she hoped because some groups had not put their problems in order on the chart paper, making it impossible to line them up when placed on the board. She felt that the goal of comparing and contrasting the problem types was not reached because the children could not *see* the relationships visually. After she was done, each teacher shared his or her observations of individual students and groups, and what they had observed about the students. During that time, the facilitators took notes. After all teachers shared, the facilitators raised questions to stimulate development of the lenses that would encourage rich discussion about the mathematics and student learning. For example, from this initial lesson they asked:

- What do you think about the difficulty level of the problems? [*Student lens*]
- Do you think the order in which the problems were solved/presented impacted student understanding? [*Curriculum developer lens*]
- Did you notice dominant students in the groups? Do you think that impacted the learning? What could we do to limit that in future group work? [*Researcher lens*]

As experienced observers of classroom interactions, the facilitators were accustomed to carefully watching and listening to students, noticing themes, and raising questions about what they observed. The questions they raised emerged from the discussion, but were not ideas that the teachers addressed directly in the discussion. The questions were designed to move the teachers beyond the limits of their

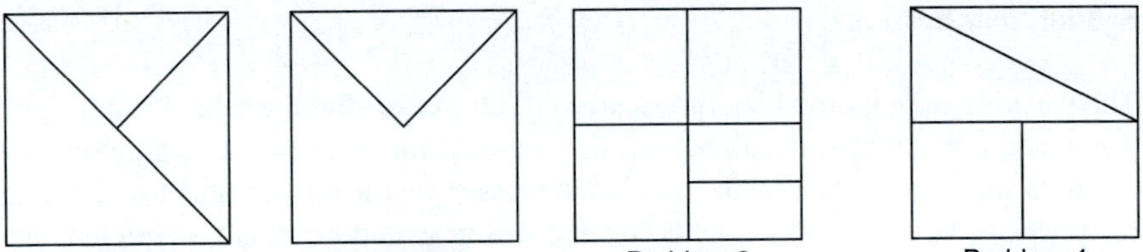

Fig. 1 Student problems for last lesson study

observations. Finally, before closing the debriefing, the facilitator asked if anyone wanted to make a comment or reflect on the *process* of lesson study. Teachers commented on how difficult it was to watch the students and not interact with them, i.e., to remain quiet. At the end, Peg made a statement that promoted the development of the *culture* of lesson study within the group. She said, "A couple of times during the discussion today I was getting defensive about what was being said, but I realized that everyone was taking responsibility for the lesson and it wasn't about me it was about the lesson. It was hard, but I think I understand better about the process and what we are trying to do."

A complete description of the three subsequent lessons is not included in this chapter. The remaining three lessons followed the same format as described above. It is, however, necessary to visualize the problems in the last research lesson in order to interpret the research findings discussed in Part 2. The final lesson required the students to name the fractional parts of a square region (see Fig. 1). The students were asked to label the fractional part of the whole that each section represented. No parts were shaded.

Part 2: The Research Project

Data Collection and Analysis

The previous detailed description of a novice lesson study community provides the context in which to understand the findings of this study. There were three subsequent lesson study cycles during the school year that followed the same sequence as the first. For the purposes of the research, we analyzed the planning and debriefing sessions for the first lesson and the last lesson.

Research Question

In order to answer our research question on whether the teachers developed the critical lenses (Fernandez et al. 2003) necessary to be a productive lesson study community, a qualitative methodology was employed using enumerative analysis

(Lincoln and Guba 1985) in which previously defined units or categories are subjected to systematic counting or identification. The first and last lesson study cycles were videotaped and supported through field notes. The videos were transcribed. Using the group as the unit of analysis, the transcriptions were coded for evidence of a researcher lens (R), a curriculum developer lens (C), or a student lens (S). Since prompting from the facilitators was used to encourage development of the lenses, only discussion *prior to* facilitator comments was coded. Both the frequency and nature/quality of the comments were noted.

For the student lens, coding was made for any comment *about* students: what they did, what they said, or what they appeared to understand. Any comment about how the lesson was constructed or ordered was coded as curriculum developer lens. For the researcher lens, we looked for initiation of questions that could be answered through experimentation similar to action research.

Results

Results from analysis of the video data show that over the course of the year the teachers developed a qualitatively richer student lens and curriculum developer lens. While comments about students were found at both the beginning and the end of the year, the substance of the comments was quite different. Teacher comments coded as the student lens from the initial lesson were primarily about *how students behaved*:

> The children were so cooperative.
> My group was very calm and respectful.
> I did notice they played with their name tags.
> Only Molly Emma played with the manipulatives in my group.

and *what the students did* during the lesson:

> They responded well.
> They listened attentively to responses of the other children.
> My boys spoke more than my girls.
> My group shared the jobs equally.
> Megan drew a picture to answer the questions. Jason was definitely in charge and Moses was happy to let him be in charge.

Only one comment stood out that suggested some deeper thoughts about student understanding. One teacher said:

> I don't think they noticed the differences in the word problems. Only one boy said "all these are about Mr. Bob."

In the final lesson on fractional parts of a square region, teachers commented on what *students appeared to understand* about the mathematics:

> It was interesting because in [problem] number two, India, if she would have, if Steve would have let her say exactly what she thought right when she saw it—which was [pointing to the diagram] "this is one-fourth, one-fourth, one-fourth" which when she looked at

> that large space, she actually *saw* that. But then Steven said, "there's no line here" and then India said "OK," when he did that, she looked at the diagram, and she goes [pointing to the diagram] "so this is one-fourth, one-fourth, one-fourth." She saw it right off the bat.... Briana... she's the one who when they were doing these examples just drew that, one-third, one-third, one-third, and so she used her diagonals.
>
> And, yeah, at the beginning when you were getting information from them, I thought it was great they said equal a lot, they said one-third because three makes a whole so that made me think—I don't know how, that child obviously knew denominator, but I don't know how.

and what was *confusing* the students:

> I still don't know if they know really what the denominator represents, though, or they're not there yet. I mean I think they know they can divide things up, but I don't think they know what that denominator really is and are able to connect that to what they're writing to what the figure is and, and what you're representing with the figure.
>
> It's hard for them to jump from the one in the numerator to anything beyond one in the numerator.
>
> I have down the word denominator in big letters because I really think that the concept of denominator is just hard. It seemed like my groups could divide anything into one-eighth, one-sixth....
>
> I think that was the big challenge of this—when they see that larger region—that in their minds they put those ones together and make it two-fourths instead of one-fourth and one-fourth.

As the quotes suggest, the teachers began to look at problems through the student's eye, acknowledging what learners were unfamiliar with or struggling with, e.g., "not having parts shaded-in" and "unequal fractional regions." They discussed how the way a problem was presented would impact student thinking. They began to unpack the mathematics in a problem and ponder about how students would approach the different parts.

Changes in the comments coded as the curriculum developer lens also changed qualitatively over the year. Initial comments focused more on how the organization of a lesson or materials *aided in management issues* so the students would stay on task or not get confused.

> I think your giving an overview of the lesson was helpful.
>
> I think turning over the bags helped so kids wouldn't play.
>
> I am not sure about the manipulatives. They just played with them.

By the last lesson, comments focused more on *how the organization of the lesson supported or hindered student understanding* and development of the concept being taught.

> I wonder about shading, if there was something... the idea that a lot of their history is with shading and the way they've learned to identify parts is by what's shaded. We had no shaded parts and they had to just label from that.
>
> And I'm thinking [problem] number two, how we could have done that one, adjusted it so it made more sense or if there was something we could have worked up to do number two, shading the one-fourth.
>
> I guess it was two things, though, I mean if we just wanted them—not shading it was trying to get them to not be familiar because they're use to shaded, but also we're giving them a large region, so maybe one or the other would have been good to do for number two.

> ...in the large group discussion could you, when you're talking about the thirds on the figure, could you shade, you know, two of those thirds, and say "OK, well what part of the circle is that?" So then that maybe they could take that piece of the discussion and transfer it to the problem.

Not only did the teachers point out specific ideas that were difficult (e.g., the use of unequal regions and the lack of any shading), but they made suggestions for how the lesson should or could be changed to scaffold understanding.

No evidence was found of the researcher lens in the first cycle. The teachers did not question their practice. In the final session, the teachers asked, "I wonder..." questions where they pondered how something was presented or understood. For example, the quote from above where the teacher wonders about the lack of shading of the diagrams:

> I wonder about shading, if there was something...the idea that a lot of their history is with shading and the way they've learned to identify parts is by what's shaded. We had no shaded parts and they had to just label from that.

We would expect that if the group of teachers held a rich researcher lens, the discussion would move on to suggesting activities or experiments to test their hypothesis. This did not happen.

Discussion

In Part 1 of this chapter we provide a case study of the development and implementation of a novice lesson study community by a group of third-grade teachers. The detailed description of each piece of the process provides a context within which to understand the significant change in the group. Adaptation of the lesson study model was necessary to meet the particular circumstances of the school system, but was done so without removing the essential pieces of the model (See Chap. 1 by Murata, this volume). This detailed account also exposes the cultural and educational differences of U.S. teachers. The teachers were not comfortable with taking the lead in planning lessons. They were not comfortable with being observed. They became defensive when their ideas were challenged. In addition, limited-to-no experience of observing other teachers' teaching and analyzing student thinking minimized the depth of initial discussions. However, as Fernandez et al. (2003) suggest, adaptations to the lesson study process are necessary for U.S. teachers to "move beyond the popularized view that currently exists in the U.S. of lesson study as a completely teacher-led and teacher-run activity" (p. 183). As this study confirms, the knowledgeable others (facilitators) can support lesson study communities in developing the critical lenses necessary to "push their lesson study practice into rich arenas" (p. 182). There is room for the active support of external facilitators who are knowledgeable about the lesson study process and who embrace the values of lesson study: A culture of self-criticism, openness to the ideas of others, and willingness to embrace mistakes.

The U.S. teachers in this study showed gains in developing two of the three critical lenses needed to create and reflect on mathematical lessons that dramatically move student learning forward. And, while their initial discussions demonstrated a stance that aligned with many standards-based instructional practices, the change in the richness of their discussions from the beginning to the end of the year suggests that merely holding beliefs that support standards-based learning environments does not guarantee the profound knowledge of planning, teaching, or learning that aligns with those beliefs.

The results from this chapter address only one small part of what we need to learn about lesson study as a useful model of professional learning. Clearly, there is still much to be learned. How do we develop the culture of lesson study with U.S. teachers? What role should the outside expert (facilitator) play in the lesson study process? How much should he or she intervene? What resources are needed to support U.S. teachers in developing the deep content knowledge and pedagogical knowledge needed in a lesson study approach to professional learning? How will participation in lesson study impact student learning? Research in this arena is still in its infancy. These questions are just a few that need to be explored as we study the lesson study model of professional learning and attempt to implement it in new arenas.

References

Cooney, T. J., & Shealy, B. E. (1997). On understanding the structure of teachers' beliefs and their relationship to change. In E. Fennema & B. S. Nelson (Eds.), *Mathematics teachers in transition* (pp. 87–109). Mahwah: Lawrence Erlbaum Associates.

Fernandez, C., Cannon, J., & Chokshi, S. (2003). A US–Japan lesson study collaboration reveals critical lenses for examining practice. *Teaching and Teacher Education, 19*, 171–185.

Lewis, C. (Producer). (2000). *Can you lift a 100 kilograms?* [videotape]. (Available from Mills College Department of Education, 5000 MacArthur Boulevard, Oakland, CA 94613).

Lewis, C. (2002). *Lesson study: A handbook of teacher-led instructional change.* Philadelphia: Research for Better Schools.

Lincoln, Y. S., & Guba, E. G. (1985). *Naturalistic inquiry.* Beverly Hills: Sage.

Lortie, D. (1975). *Schoolteacher: A sociological study.* Chicago: University of Chicago Press.

Ma, L. (1999). *Knowing and teaching elementary mathematics: Teachers' understanding of fundamental mathematics in China and the United States.* Mahwah: Lawrence Erlbaum.

National Council of Teachers of Mathematics. (2000). *Principles and standards for school mathematics.* Reston: National Council of Teachers of Mathematics.

Pajares, M. F. (1992). Teachers' beliefs and educational research: Cleaning up a messy construct. *Review of Educational Research, 62*(1), 307–332.

Richardson, V. (1996). The role of attitudes and beliefs in learning to teach. In J. Sikula (Ed.), *Handbook of research on teacher education* (pp. 102–119). New York: Simon & Schuster.

Stigler, J., & Hiebert, J. (1999). *The teaching gap: Best ideas from the world's teachers for improving education in the classroom.* New York: Free Press.

Influence of Lesson Study on Teachers' Mathematics Pedagogy

Jo Clay Olson, Paul White and Len Sparrow

Lesson study enables teachers to use their desire to understand their students' learning on a journey of self-discovery. In lesson study, teachers set their own goals, create a research lesson, investigate student learning, and interpret students' responses from their own perspectives and knowledge. This process provides teachers with opportunities to collaborate, reflect, and analyze student responses. Thus, lesson study encourages teachers to develop a continuous practice of inquiry in which they investigate student thinking and refine their own practice.

Teachers have multiple opportunities to gain insights about teaching and learning while participating in lesson study. These insights are seen as valuable by teachers (Fernandez and Yoshida 2004; Lewis et al. 2004; Southwell and White 2004; Wiburg and Brown 2007). But, do teachers integrate these insights into their daily practices? Specifically, two questions were posed:

1. In what ways do teachers integrate insights into classroom practice gained from lesson study professional development?
2. What aspects of lesson study promote teacher reflection, collaboration, and change in classroom practice?

Ten teachers joined one of three lesson study groups during the beginning of the school year. Two groups completed their lesson study and one group abandoned it while planning their research lesson. From this context, it was possible to contrast the learning of teachers who completed a research lesson with those that did not and identify aspects of lesson study that supported teachers' professional development. In this paper, reasons that led the one team to abandon lesson study are outlined. In addition, factors that both encourage and discourage teachers' involvement in the lesson study process are noted.

J. C. Olson (✉)
Washington State University, Pullman, WA, USA
e-mail: jcolson@wsu.edu

L. C. Hart et al. (eds.), *Lesson Study Research and Practice in Mathematics Education*,
DOI 10.1007/978-90-481-9941-9_4, © Springer Science+Business Media B.V. 2011

Context of Lesson Study

Over 200 elementary teachers and administrators in a Midwestern school district sought support from the mathematics education faculty in a large university as they phased in *Investigations in Number Data, and Space* ([*Investigations*], Akers et al. 1997), a reform elementary mathematics curriculum that was aligned with both state and national standards. The goal of the collaboration between mathematics educators and the school district was to increase student achievement in mathematics by improving the district's K-5 mathematics program and grade-to-grade coordination through a three-year grant (PRIME).

The teachers participated in PRIME by attending one-week summer institutes, monthly professional development sessions during the school year, and coaching sessions in their classrooms to discuss pedagogy. The teachers requested that the leadership team create a framework for continued professional development that supported mathematics reform after PRIME. A Leadership Institute was created in response to these concerns and all the teachers participating in the project were invited to submit an application to join.

The Leadership Institute provided teachers with the skills needed to facilitate discussions about the teaching and learning of mathematics and promote educational reform. Ten teachers representing different elementary schools volunteered to participate in the Leadership Institute and also to facilitate professional development sessions for teachers in their school.

The institute was broken into three segments. Segment 1 was a summer institute which met for one week in the summer. Teachers became familiar with research on leadership, learning, and different models of professional development. The teachers selected lesson study as a professional development model after watching the video, *Can You Lift a 100 kilograms* (Lewis 2000). They believed that lesson study would allow them to investigate student cognition in a meaningful way and serve as a model that would engage their colleagues. Three lesson study teams formed along grade level bands. Teachers engaged in lesson study during segment 2. Monthly meetings were held after school for six months in which teachers planned, taught, and reflected on a research lesson. Segment 3 spanned the next 12 months. It focused on developing teachers' skills, planning professional development, and facilitating discussions with peers.

Theoretical Framework

One characteristic of lesson study is teacher collaboration. Within a collaborative group, individuals interact with each other through words, gestures, and written symbols. Symbolic interactionism is a perspective from which researchers can interpret how interactions between people create meaning through observations, experiences, and words (Kuwabara and Yamaguchi 2007). Knowledge grows out of interactions between people and it is the interplay between personal and social meanings that creates new ideas. This knowledge is dynamic and changes as individuals create and reconstruct it. The character of verbal language changes as people assimilate new ideas (Kumpulainen and Mutanen 2000).

From the perspective of symbolic interaction, examining the teachers' discourse provides a window into their thoughts. Boden (1990) and Kumpulainen and Mutanen (2000) found that changes in classroom discourse indicated a shift in a teacher's interpretation of reform mathematics. The discourse among teachers during the Leadership Institute and between each teacher and her students during classroom observations were analyzed for new patterns of interaction. Thus, changes in the teacher's discourse were interpreted as a change in pedagogy.

Methods

To describe how lesson study influenced teachers' practices, a research design using case study was created (Merriam 1998). Data collected included (a) video recordings of instruction, (b) audio recording of instruction and lesson study meetings, (c) interviews, (d) teacher reflections, and (e) field notes of classroom observations and Leadership Institute meetings. Transcriptions were made of the recordings. Data were analyzed using a qualitative program (winMAX, Kuckartz 1998) that allowed hierarchical coding to characterize teachers' practices before beginning, during, and after lesson study. Data analysis was compressed using conceptual and time-ordered matrices to identify patterns that indicated change of practice (Miles and Huberman 1994). These changes were characterized and compared to lesson study activities to describe how the lesson study activities influenced teachers' growth. Observations continued for an additional 12 months to determine whether changes in teachers' practices were sustained.

The purpose of each research lesson was to create opportunities in which teachers could ask students questions that would uncover misconceptions and identify the point when a problem-solving strategy led students to an erroneous solution. Two frameworks, namely the tasks' cognitive demand (Stein et al. 2000) and secondly question types (Driscoll 1999), were used to create problems and hypothetical questions for the research lesson that would engage students in the exploration of mathematical ideas and the articulation of their thinking.

Case Study Teachers

Five elementary teachers (Table 1) who were members of the Leadership Institute volunteered to participate in a case study. The other five teachers in the Leadership Institute participated as team members and an analysis of their practices was not investigated.

Table 1 Case study participants in the Leadership Institute

Participant	Prior experience	Characteristics
Place Value lesson study (three teachers in the team)		
Pat grade 1	1st and 4th grade for 12 years	• Passionate about deepening her mathematical content knowledge • Enjoyed discussions about mathematical thinking • Principal was surprised at her interest in math and leadership
Paula grade 2	1st and 2nd grade for 16 years	• Limited to procedural knowledge • Hoped to gain insights about children's problem-solving approaches • Principal designated her as a model for literacy instruction
Count-Back-Change lesson study (three teachers in the team)		
Courtney grade 3	K and 3rd grade for 30 years	• Limited mathematical understanding • Wanted to ask new kinds of questions • Principal provided limited support
Subtraction Regrouping lesson study → Curriculum Guide (four teachers in the team)		
Rachel grade 4	4th grade for 15 years	• Conceptual and procedural knowledge was built from science applications • Wanted to expand her expertise in math • Principal gave her extensive praise for her dedication and teaching
Rose grade 4	3rd and 4th grade for 20 years	• Procedural knowledge was connected to different representations • Wanted to meet the wide range of students' abilities in mathematics • Principal recognized her as an outstanding teacher • Granted release time and rarely taught mathematics

Findings and Discussion

Two phases are used to describe how lesson study influenced teachers' pedagogy. First, the initial lesson study experience for the three teams of teachers is described. Second, the lesson study experience of the two teams who completed their research lesson is outlined followed by a description of the group who abandoned their research lesson. These descriptions provide the context for contrasting the pedagogy of teachers who completed their research lesson with those who did not. The changes are characterized through illustrative examples and then linked to specific aspects of lesson study that supported teachers' professional development.

Initial Lesson Study Experience

The three lesson study teams began their lesson studies simultaneously at the beginning of the new school year with the goal of uncovering students' conceptions about a mathematical idea to help teachers develop a practice of inquiry. This required teachers to shift their focus from step-by-step teaching actions to the priority of student thinking. The teachers were familiar with professional development in which they met to discuss mathematical concepts that students struggled with and then planned instruction that would support student learning. The lesson study experience went beyond their previous experiences as a catalyst for teachers to become coresearchers with the facilitator to investigate student thinking.

At the second lesson study meeting, teachers were confronted with a choice: becoming a coresearcher or being a recipient of research. Two of the lesson study teams decided to investigate students' mathematical thinking as coresearchers and the other team decided to create a 4th-grade curriculum plan. Reasons for these choices are discussed after a description of the two teams that completed their research lesson.

Completing the Research Lesson

Three months later, a teacher in the Place Value (1st/2nd grade) and Count-Back-Change (3rd grade) volunteered to teach the research lesson. The problem that the 1st/2nd-grade team used was, "There are 36 cupcakes. We want to put 10 cupcakes in each box. How many boxes would we need to fill? How many extra cupcakes?" This problem allowed them to investigate how young students thought about place value. The problem that the 3rd-grade team asked was, "Use a count-back-change method to find the change for a snack that costs 65 ¢ if you give the clerk one dollar." Their goal was to investigate how students made change, but they immediately began instructing students how to use the count-back-change strategy.

During the research lesson, the teachers moved around the classroom to observe what students were doing at their desks and asked the students questions to find out why they used a particular approach to solve the problem. The interaction here was a deliberate strategy to help teachers gather more close-up evidence of student thinking, providing a necessary experience enabling teachers to focus on student learning during a lesson observation. After observing the lesson and interacting with students, the teachers met with the facilitator for debriefing.

The Place Value and Count-Back-Change teams discussed the difficulties that students had using manipulatives to solve the problems designed for the research lesson. The facilitator recognized that it was difficult for the teachers to talk about students' learning and misconceptions without immediately creating an intervention to remedy their limited understanding. After the initial debriefing, she selected a portion of the videotapes and transcribed the interactions for each team to prompt deeper reflection. The teachers' in each team watched their videotape and classified the types of questions that they asked. Through this analysis they realized that many of their questions were "leading" the students toward a preferred solution strategy rather than uncovering students' conceptions related to the content. As a result, the Place Value team decided to revise and reteach the lesson to further investigate student thinking. The Count-Back-Change team shifted their goal from investigating student thinking to designing a lesson to teach a particular counting procedure. Thus, reteaching of the lesson was not considered worthwhile by the team because the purpose of the lesson had changed.

This change of goals created a set of challenges and discomfort for the Count-Back-Change team. The inconsistency between wanting to let students use their own strategy and believing in the usefulness of a preferred procedure created tension. This clash between new knowledge and long-term beliefs is felt by many teachers as they adapt mathematics reform recommendations into their pedagogy. The Count-Back-Change team wanted to follow the reform recommendations, but they found themselves using the pedagogy of lecture, telling, and showing students a particular procedure that seemed disconnected to students and their ways of thinking.

The Count-Back-Change team struggled to resolve their new knowledge with a set of beliefs built on a history of experiences. The teachers preserved the procedure that did not make sense to students. Lesson study helped them notice the tension and created an opportunity to examine it. The 3rd-grade teachers discovered a conflict between knowing and believing. In spite of not resolving the conflict, the experience prompted Courtney to change her pedagogy by adopting an inquiry stance while working with her students.

Outside Japan, lesson study frequently takes different forms, revising and reteaching a lesson is an example of this variation. In this study, the choice of whether to repeat the lesson was left to the two teams. There are multiply ways that lesson study can be enacted to meet the specific needs of teachers and professional development goals.

Abandoning the Research Lesson

The 4th-grade teachers wondered why so many of their students did not remember how to regroup when subtracting and formed a team to investigate regrouping. In contrast to the other two lesson study teams, these teachers were ready to teach a lesson on regrouping at the end of the first lesson study meeting. The facilitator intervened and suggested they give a problem (Fig. 1) to their students and gather

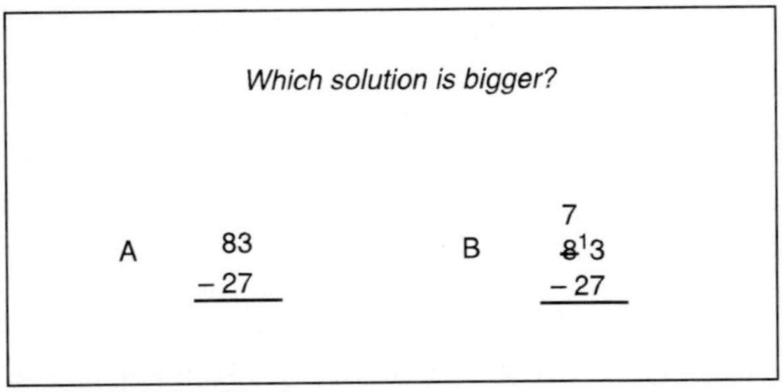

Fig. 1 Suggested problem for 4th-grade students to consider

students' responses. The teachers discussed the problem and anticipated that many students would say that problem A was bigger than B. The facilitator asked why they thought that and the teachers responded that students would subtract the tens without regrouping.

One vocal teacher (Rose) became quiet and withdrawn while the other teachers talked about why students might subtract the tens without consideration of the units. The withdrawal of Rose signaled that she did not value the discussion, and investigating student thinking was a diversion from her goal: design a lesson to teach regrouping. The session ended with three other teachers agreeing to pose the problem to their students and collect their responses for the next meeting.

The next lesson study planning meeting began with Rose forging ahead with planning the regrouping lesson. The facilitator again intervened and asked how students responded to the problem. Rose spoke first, "We didn't ask them. We knew what would happen and we didn't have time." The facilitator reemphasized the importance of taking advantage of this collaborative time to investigate why students forgot the procedure. The team finished planning their lesson to teach regrouping by the end of the session.

These 4th-grade teachers felt pressure to "get the kids ready for the state test" and creating a lesson to reveal students' conceptions was moving too slowly. They sought quick solutions to improve students' test scores on the mandated test. Investigating students' conceptions was valued if it could be done quickly and a remedy was available to "fix it." Simply put, lesson study took too much time, distracted them from "getting through the curriculum," and managing the three sets of materials that they had to work with (*Investigations,* textbook, and supplemental activities). They abandoned lesson study and began organizing the materials into a curriculum guide.

Two months later, the facilitator noticed that many of the activities placed in the guide by the 4th-grade teachers could be described as "drill and practice." She asked the team to analyze the tasks using the task's cognitive demand (Stein et al. 2000) and determine whether each task met the projects' objective, to develop the mathematical thinking of students, before including the activity in the curriculum plan. The cognitive demand of tasks characterizes problems in four ways: *Memorized* (reproducing previously learned facts, rules, or definitions); *Procedures Without Connections* (using a specific procedure without explanation); *Procedures with Connection* (linking procedures with conceptual ideas); *Doing Mathematics* (analyzing the problem and constraint to create solution strategies from which mathematical conversations emerge). Tasks that rely on memorized information have the lowest level of cognitive demand while tasks that require students to do mathematics have the highest level of cognitive demand.

To clarify the levels of cognitive demand, a page from the textbook was selected by the teachers. The facilitator made the mathematical concepts in the lesson explicit and classified it as a procedure without connections. Problem solving was not evident and it did not represent the kind of activities that met the goals for the professional development project. The facilitator encouraged the 4th-grade team to use the cognitive demand framework as a filter before including any activity in their curriculum guide.

After a short discussion, the team characterized the next activity as a task with high cognitive demand even though it was similar to the lesson just discussed. Rose explained, “Students [can be asked to] describe a real life situation for which a given mathematical equation could be used to solve the problem.” Rachel continued, “This type of task has high cognitive demand because there are multiple answers and it requires complex and non-algorithmic thinking.” Rose added, “It develops students’ mathematical thinking because it required students to solve their own problem.” The teachers decided that all of the textbook pages could be classified as high cognitive demand by asking students to create a situation for one of the exercises. Thus, the team avoided the opportunity to critically consider the type of learning opportunity created by the lessons.

Teachers’ Pedagogies

The five teachers who began lesson study were dedicated and thoughtful about instructional decisions and students in their classrooms performed well on standardized tests. They asked questions that led students to expected responses and their discourse followed a traditional IRE pattern in which the teacher initiates a question, listens to a response, and evaluates the response (Cazden 2001). Thus, the teachers’ pedagogies were similar (Table 2).

The pedagogies of Pat (Place Value lesson study) and Rachel (Regrouping/Curriculum Guide team) are now described to provide a more detailed look at how lesson study and the ensuing collaboration can result in two totally different professional learning experiences.

Table 2 Initial pedagogies of the case study teachers

Teacher	Questioning	Discourse pattern	Task’s cognitive demand	LS participation
1st/2nd grade: Place Value lesson study				
Pat	Leading[1]	Traditional IRE	Procedures with and without connections	Fully
Paula	Leading	Traditional IRE	Procedures with and without connections	Fully
3rd grade: Count-Back-Change lesson study				
Courtney	Leading	Traditional IRE	Procedures with connections	Fully
4th grade: Regrouping lesson study → Curriculum Guide				
Rachel	Leading	Traditional IRE	Procedures with and without connections	Withdrew
Rose	Leading	Traditional IRE	Procedures with and without connections	Withdrew

[1] Questions that led students to an expected response

Initial Pedagogies of Pat and Rachel

Prior to beginning lesson study, Pat and Rachel had many similar practices. They mainly designed lessons to help students learn procedures which resulted in teaching that required a low cognitive demand. Both were confident teachers and espoused similar beliefs not necessarily reflected in what happened in the classroom. Pat stated, "I present situations to students, question them, and let them explore." Rachel claimed she pushed her students "beyond finding an answer and stopping. We try to find if there are other possible answers, why or why not."

Pat's Initial Pedagogy

Pat's 1st-grade mathematics instruction was divided into two segments, calendar math and class instruction. At the beginning of the project, Pat's pedagogy was characterized by the following three qualities: (a) questions were asked to help learn procedures for describing the passage of time during a calendar math, (b) questions were asked for helping students understand the procedure to solve problems during class instruction, and (c) students were given time to respond to questions.

During calendar math, a child put one straw to a jar to represent each school day. When ten straws were collected in the jar marked ONES, they were bundled with a rubber band and placed in a container marked TENS. The following excerpt from October is typical of Pat's calendar math and provides an illustrative example of her questioning.

153: Pat	How many days have we been in school?
154: Jeffrey	Thirty-nine.
155: Pat	Thirty-nine, very good. How many ones do we have?
156: Ellen	Nine.
157: Pat	How many ones are in 39?
159: Brandon	Nine.
160: Pat	Nine. How many tens do we have? Chris
161: Chris	Ten.
162: Pat	What was my question?
163: Chris	Three.
164: Pat	What was my question?
165: Chris	(Pause 4 seconds) How many tens is there?
166: Pat	Yes, how many tens do we have?
167: Chris	Three.

Pat began mathematics lessons with calendar math and she asked the same set of questions (lines 153, 155, 157, 160). The cognitive demand was low as students were focused on reproducing an expected response. In this example, Pat elevated the task to *Procedures with Connections* when she asked the student to restate the question (line 164) before she accepted his response (line 167). Throughout this episode and other interactions, Pat maintained control of the classroom discourse using the traditional IRE discourse pattern.

Rachel's Initial Pedagogy

Rachel's 4th-grade mathematics instruction was divided into two segments, warm up and problem solving. At the start, Rachel's pedagogy was characterized by three qualities: (a) questions were asked to make connections between representations and mathematical symbols, (b) questions during class instruction helped students create and practice definitions, and (c) students were given time to respond and self-correct themselves. She typically presented a task to elicit students' prior knowledge at the beginning of each mathematics lesson. The following excerpt illustrates both her introduction and the interactions that she typically had with students.

Rachel held up several rectangles that represented halves of a square. Students taped two rectangles together to form a square that represented one whole. As they taped the rectangles together, Rachel asked a series of questions that led the students to a correct symbolic representation.

28: Rachel	How many halves are here (pointing at one rectangle)?
29: Marie	One.
30: Rachel	And here (pointing a rectangle next to the first rectangle)?
31: Marie	One.
32: Rachel	How many altogether?
33: Marie	Two.
34: Rachel	Can you write 2 halves? (Pause 5 seconds) The whole is divided
35	into how many parts? (Pause 4 seconds) Who can help?
36: Steven	Two.
37: Rachel	Two, that's the denominator. And what is the numerator?
38: Mark	Two?
39: Rachel	That's right. Can you write that (to student 1). That's right, put a 2
40	for the denominator and a 2 for the numerator.

Rachel had selected an activity that provided an opportunity for her to assess students' knowledge and review definitions or procedures. The cognitive demand was low as students were focused on reproducing an expected response (lines 29, 30, 32). When Marie was unable to write a symbolic representation of halves, Rachel asked another student to answer her question (lines 35 and 37). Rachel then led Marie to a correct symbolic representation of two halves. Like Pat, Rachel maintained control of the classroom discourse using a traditional IRE discourse pattern.

Pedagogies After Lesson Study

The three case study teachers who completed their lesson study changed unique aspects of their pedagogy (Table 3). These changes were revealed in their interactions with students and colleagues.

In contrast, the facilitator did not discern changes in the 4th-grade teachers' pedagogy who did not complete the lesson study process. Thus, only one excerpt

Table 3 Pedagogies after lesson study

Teacher	Grade level	Classroom questioning	Discourse pattern	Task's cognitive demand
Place Value lesson study				
Pat	1	Elicit mathematical thinking[1]	Reform: Student Led Discussions[2c] (Initially teacher dominated)	Procedures with connections and doing math
Paula	2	Prompt reflection[1]	Reform: Accept Alternative Strategies[2a] (Initially teacher reinforced specific strategies)	Procedures with connections
Count-Back-Change lesson study				
Courtney	3	Elicit mathematical thinking	Reform: Community of Inquiry[2b] (Initially teacher had sole authority)	Procedures with connections and doing math
Regrouping lesson study → Curriculum Guide				
Rachel	4	Leading	Traditional	Procedures with and without connections
Rose	4	Leading	Traditional	Procedures with and without connections

[1] Questions that elicit mathematical thinking encourage students to make observations about patterns and create conjectures. Questions that prompt reflection encourage students to reflect, justify, and extend their thinking (Driscoll 1999)

[2] Three types of discourse indicate reform pedagogy. They include (a) Accepting Alternative Strategies where students present their solutions and provide a justification, (b) Community of Inquiry where students pose problems and initiate topics for investigation, and (c) Students Led Discussions (Cazden 2001)

from Pat's classroom interaction is presented to serve as illustrative examples of the teachers' evolving pedagogies.

Changes in Pat's Pedagogy

Pat's pedagogy changed in three ways. She shared responsibility for leading the class discussion with children, she elevated the cognitive demand of the task during instruction, and she wrote the entire instructional task on the board instead of a short version that displayed only the important information. An example of how she shared authority with a child is presented as an illustrative example of the changes she made. Pat drew a tic-tac-toe grid on the board to represent a portion of a hundreds chart and placed a 23 in the center box (Fig. 2).

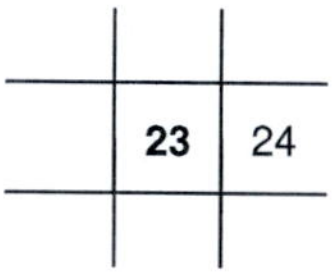

Fig. 2 Tic-tac-toe grid drawn by Pat (*bold*) and Jasmine's response

Pat gave the marker to Jasmine. Jasmine wrote the number 24 in the square to the right of the 23. She then turned and faced the class.

324: Jasmine	Do you have a question? Raise your hand.
325: Student 1	How do you know that?
326: Jasmine	I counted, after 23 comes 24.
…	
332: Student 3	How much tens is there in 20?
333: Jasmine	Uhm, the amount…(pause 14 sec)
334: Pat	Did you understand her question? She asked you how many tens
335	there were in 20.
336: Jasmine	There's 20.
337: Pat	How many did you say?
338: Jasmine	Two.

Jasmine assumed responsibility for the classroom discourse and asked the class if they had any questions (line 324). One student asked her to justify the placement of 24 (line 325) and another child asked her to think about the meaning of twenty (line 332). During the interaction between students, Pat assisted her students as they developed their questioning skills (lines 334). This classroom discourse pattern was a departure from the traditional pattern with the teacher maintaining control. Pat shared authority with students by allowing them to lead the discussion and pose questions to each other.

Summary of Teachers' Pedagogy After Lesson Study

Detailed analysis of the three case study teachers who completed the lesson study process changed their pedagogies over 18 months. Pat, Paula, and Courtney asked more purposeful questions and posed contradictions based on students' responses. Before lesson study, the three teachers maintained control of the classroom discourse and frequently asked students "Why?" without knowing what to do with the information that they gained. After completing their lesson studies, these three teachers asked purposeful questions and used the students' responses to guide instructional decisions. Thus, we describe these changes as a practice of inquiry in which they asked questions, listened to students' responses, analyzed those responses, and made instructional decisions.

In contrast, Rose and Rachel showed students procedures and attempted to help students understand them. Rachel continued to introduce a lesson by asking students to recall previously learned information and suggesting procedures to follow. The suggested procedures were further explored using manipulatives to concretely demonstrate the procedure. The classroom discourse remained traditional. Rachel asked questions that led students to expected responses and reinforced their explanations by repeating students' responses. Rachel viewed her primary job as preparing students for the state mandated test by teaching them to replicate preferred procedures.

Linking Teacher Change to the Lesson Study

Examining the pedagogy of those who completed the lesson study allows a better understanding of what aspects of lesson study supported sustained changes in pedagogy. First, the experiences of the Place Value team are discussed and linked to changes in Pat's and Paula's pedagogy. Second, the experiences of the Count-Back-Change team are discussed and linked to changes in Courtney's pedagogy.

Place Value Team

During the first lesson study meeting, the 1st-grade teachers theorized that students who understood place value would break a multi-digit number into groups of tens and ones and utilize the structure of the hundreds chart to solve problems. They indicated that most of their students did not break multi-digit numbers apart and relied on the inefficient method of counting by ones to solve problems. During the second meeting, the primary teachers discussed how pictures could be incorporated with the hundreds chart to develop place value understanding.

They selected a task that connected a picture with the place value and created the following two problems: "There are 36 cupcakes. We want to put 10 cupcakes in each box. How many boxes would we need to fill? How many extra cupcakes?" and "We have three boxes of cupcakes and 6 extra on a plate. Each box has 10 cupcakes. How many cupcakes do we have?" During the third meeting, these primary teachers discussed how teachers maintain or reduce the cognitive demand of tasks by the types of questions that might be asked. This discussion led to the creation of probing questions for the research lesson that would not reduce the cognitive demand of the problems.

The Place Value team finalized their research lesson during the fourth meeting. They decided to state the problem orally to the students and then Pat would write the important information on the board, "36 cupcakes" and "10 in a box." After a lengthy discussion about whether students needed the shortened problem, Pat concluded, "I think I inadvertently lower[ed] the cognitive demand [of the problem when writing the short version on the board]." Her critical reflection created an opportunity for team members to discuss their classroom routines and assumptions that impacted a task's enacted cognitive demand. During this discussion, the teachers shifted from planning a single lesson to sharing aspects of their own pedagogy that influenced students' opportunities to learn. This shift of focus from the impersonal to personalized reflection on their own actions and beliefs can be thought of as critical reflection. It goes beyond reflection on actions to deeply considering how those actions reflect a set of beliefs and assumptions that may not in fact support student learning.

After this critical reflection, Pat maintained the cognitive demand of posed tasks by writing the full problem on the board instead of an abbreviated form. She posed tasks with higher cognitive demand during calendar math and established a new discourse pattern during a hundreds chart activity by encouraging students to lead a classroom discussion.

Paula shifted the focus of her reflections. She initially interpreted her students' responses as correct or incorrect. After sharing aspects of how she also reduced the cognitive demand of tasks, Paula began to speculate about her students' responses as an indication of their mathematical thinking. This was very difficult for her because she often did not know when her students responded with an important mathematical idea. She began to build her content knowledge by discussing students' responses with other teachers.

Count-Back-Change Team

During their reflection of lesson study, the intermediate teachers in the Count-Back-Change team gained a new insight when they recognized that they could not explain why a student incorrectly added a dime to 95 ¢ to make a dollar. This realization led to a discussion of questions. Questions could be posed in such a way that revealed students' thinking. More importantly, they realized the limitations of quickly determining whether a response was correct or incorrect without first considering the reasoning that the student used. Thus, they began to ask follow up questions that probed students' thinking a bit more before providing the preferred procedure.

After the analysis of the research lesson, Courtney maintained the cognitive demand of tasks when students struggled. She began to wonder why students made particular responses and asked probing questions to help her understand what they were thinking. Courtney asked her students questions to uncover their intuitive approach and then used leading questions to help them solve the problem using their own strategy. This was a surprising change because Courtney had a limited understanding of mathematics. She seemed comfortable that her students might "know more math" through their intuitions than she did. Courtney expressed joy that she was learning from them. She reflected, "I know when they get the answer right, but I learn when I ask questions to help me understand what they did. We are learning together."

Implications

This chapter describes the pedagogical changes that teachers made while engaged in lesson study. Before participating in lesson study, the teachers focused their attention on teaching methods and content delivery. They made cosmetic changes to individual tasks and lessons. Their questions led students to reaching a teacher-preferred procedure or strategy. The teachers' reflections were general and focused on delivery of the curriculum.

The three case study teachers who completed their lesson studies became more curious about students' thinking and understanding. They asked students questions to help them identify which mathematical ideas the students were applying and to understand the reasons for these choices. These teachers recognized that their actions influenced not only how students learned but also what they learned, suggesting a new insight. This insight was personal and led them to reexamine their beliefs about teaching and learning, leading to a practice of inquiry. The reflection was critical because there was some discomfort the first time they considered whether their actions supported their beliefs. Critical reflection appeared to be a catalyst for their professional growth and is typical in lesson study (Perry et al. 2002). Their insights translated into sustained change in their pedagogy, personal beliefs, and assumptions during the two years following completion of the lesson study.

In contrast, the teachers who abandoned the research lesson midstream did not change their pedagogy. They shied away from opportunities to reflect deeply on student learning and how their actions might influence opportunities for students to learn while designing the research lesson and working on their curriculum plan. Their beliefs and assumptions did not change and they continued to teach using a comfortable pedagogy of past practice.

Leadership in Lesson Study

The contrast between the teams that completed lesson study and the one that abandoned it is stark. Why might the same professional learning opportunity affect two teams of teachers so differently? One possibility is the contrasting group dynamics of the teachers who completed lesson study and those who abandoned it. How the teachers interacted with each other influenced the nature of their collaboration, decision-making process, and actions. Portes (1998) and Putnam (2000) hypothesized that individuals' social capital impacts their interactions and the dynamics of the group. Social capital refers to connections within and between social networks. While there are a variety of definitions, social capital describes an individual's level of influence derived from cultural factors such as education, social interactions, and network of relationships.

Shared Leadership

All the teachers who completed lesson study had a similar level of social capital. None of these teachers had a high-profile leadership role prior to the Leadership Institute and they were not identified as an outstanding teacher by their principal. They viewed each other as equals with different experiences from which they could learn. Thus, no single teacher in the team assumed leadership or dominated the decision-making process.

The needs of the teachers to learn and collaborate took priority in all of their actions. The teachers in both lesson study teams shared their classroom practices and the tensions between practice, knowledge, and beliefs. In doing so, they became vulnerable and found new strength from each other. Gaining strength from each other is characteristic of teams who face challenges together. Both lesson study teams voiced a need for new pedagogy because "students are not learning enough using the old methods. We can't expect to get different results if we do the same thing." With limited confidence in their initial pedagogy, the teachers who completed lesson study wanted to collaborate, learn, and share leadership with each other. They gained social capital through the process of collaboration and developed a strong network of colleagues for support. This ultimately garnered them increased influence in the school district.

Individual Leadership

In contrast, the team that did not complete their research lesson relied on a single leader for guidance. Fourth-grade was the first year that students were assessed in mathematics on a state mandated test and these test scores determined whether the school was labeled satisfactory or unsatisfactory. Thus, the teachers felt pressured to have the students ready for the test and the use of time was critical. Diversions that consumed valuable class time for work that could be accomplished more efficiently by teaching procedures was not valued. As a result, the lesson study team looked for guidance from a respected teacher.

This teacher had considerable social capital that was derived from three sources. First, she was highly respected by her principal and awarded special privileges after creating a school district writing program. Second, with confidence in her pedagogical content knowledge, she articulated content that needed to be retaught each year and anticipated students' response to new content. Third, she became a 4th-grade spokesperson who questioned how reform recommendations were translated into pedagogy. The teacher used her social capital to influence her colleagues by encouraging them to abandon the lesson study process. This process was slow and required teachers to rethink teaching and student learning. Wanting a useful product (curriculum guide) that would minimize instructional planning during the school year, the leader used her social capital to influence her team.

While the team had opportunities to think more critically about how activities may promote students' mathematical thinking, the teachers stayed focused on the curriculum guide. Critical thinking about how teachers create or limit learning opportunities is essential to develop a practice of inquiry that enhances students' mathematical learning.

The Influence of School Administration

After the Leadership Institute ended, a new superintendent was hired who refocused the school district's energy back to reading. School district support for mathematics education evaporated. The commitment of several teacher leaders kept the vision of teachers asking mathematical questions, listening to students, and planning instruction to develop students' thinking alive within their classrooms, but they were hesitant to step forward. They felt silenced by the new administration.

The administration changed several times over the next six years. The teachers who developed a practice of inquiry created networks in mathematics and extended them to support literacy education during an administration that focused attention on literacy. Eventually, the teachers who completed lesson study found a voice in the school district's mathematics instruction. Their opinions were sought when the school district revised their grade level mathematics goals and selected new supplemental materials and textbooks for the school district. In addition, they offered mathematics professional development courses for their colleagues using a workshop model.

The teachers who did not complete the lesson study process maintained their level of influence in writing and science education, but did not extend their influence to mathematics. This suggests that they maintained old networks but did not create new ones that extended their influence.

These findings indicate that lesson study supported teachers' development in two spheres. First, participation in lesson study encouraged teachers to critically reflect on their own pedagogies and this helped them develop a practice of inquiry. Classroom practices were transformed after teachers discussed a new insight with a colleague. The moment of insight was sometimes uncomfortable as teachers realized that their actions did not necessarily portray spoken beliefs. Thus, it appears that pedagogical changes are supported by a social group to create new meaning and to personalize shared experiences through critical reflection. A supportive group, such as a lesson study team, appears to encourage teachers to consciously investigate their own practices.

Second, the lesson study teams allowed teachers to establish a new network of support throughout the school district. This network enabled them to survive an administration that subdued teacher leadership through top-down decision-making policies. After several administrative changes, the teachers who completed lesson study utilized their social capital to reestablish a leadership team that supported mathematics education. Thus, it appears that participation in lesson study developed teachers' social capital through which they maintained relationships and hope for a future in which attention for mathematics education reemerged.

Creating a Successful Lesson Study

Lesson study provides a sound structure for professional learning and is certainly worth utilizing in school settings. But, there are factors which can limit its success. Teachers often work in isolation and do not share their classroom experiences. Limited venues for exchanging ideas, practices, and beliefs can bring discomfort to teachers when they try to break down the walls of isolation. Lesson study often builds a closely knit community in which teachers share experiences and vulnerabilities from which professional learning and pedagogical change can emerge.

Successful lesson study teams appear to function through collaboration in which all the teachers have a similar level of social capital. They are more likely to recognize that each person contributes to the process in different ways, but no single contribution is considered more valuable. The members of the lesson study teams that completed their research lesson appreciated the gift of time to invest in the process of thinking deeply about a lesson and student thinking. They also released themselves from external expectations to meet state mandated benchmarks so that they could modify pacing guides to work on specific content beyond the allotted time. They were open to suggestions from an outside facilitator even though at times she sometimes pushed them to think about uncomfortable things. They trusted each other and engaged in critical reflection believing that the risks they took would lead to important professional growth.

The current environment of high-stakes-mandated tests creates many tensions for teachers who want to investigate new pedagogies and consider how they influence students' learning. These tests put pressure on teachers to prepare for the exam, and schools sometimes rely on quick fix methods to make sure that students meet grade level benchmarks. Time to investigate student thinking and confidence to reallocate time by deviating from pacing plans takes courage. School administration can support teachers' professional growth and student learning by creating time in pacing guides and space in the curriculum. More research is needed to describe how lesson study and other types of professional development supports sustained teacher change and the impact of new pedagogies supported by professional development on student learning.

References

Akers, J., Battista, M., Berle-Carman, M., Clements, D., Economopoulos, K., Nemirousky, R., et al. (1997). *Investigations in number, data, and space curriculum series*. Palo Alto: Dale Seymour.

Boden, D. (1990). People are talking: Conversation analysis and symbolic interaction. In H. S. Becker & M. McCall (Eds.), *Symbolic interaction and cultural studies* (pp. 244–274). Chicago: University of Chicago Press.

Cazden, C. (2001). *Classroom discourse: The language of teaching and learning*. Portsmouth: Heinemann.

Driscoll, M. (1999). *Fostering algebraic thinking: A guide for teachers grades 6–10*. Portsmouth: Heinemann.

Fernandez, C., & Yoshida, M. (2004). *Lesson Study: A Japanese approach to improving mathematics teaching and learning*. Mahwah: Lawrence Erlbaum.

Kuckartz, U. (1998). winMAX: Scientific text analysis for the social sciences. Thousand Oaks, CA: Sage.

Kumpulainen, K., & Mutanen, M. (2000). Mapping the dynamics of peer group interaction: A method of analysis of socially shared learning processes. In H. Cowie & G. van der Aalsvoort (Eds.), *Social interaction in learning and instruction: The meaning of discourse for the construction of knowledge* (pp. 144–160). New York: Pergamon.

Kuwabara, T., & Yamaguchi, K. (2007). An introduction to the sociological perspective of symbolic interactionism: Herbert Blumer's perspective revisited. *Journal of Economics and Sociology Kagoshima University*, *67*, 1–9.

Lewis, C. (2000). *Can you lift 100 kilograms?* Oakland: Department of Education, Mills College.

Lewis, C., Perry, R., & Hurd, J. (2004). A deeper look at lesson study. *Educational Leadership, 61*(5), 18–22.

Merriam, S. B. (1998). Qualitative research and case study applications in education. San Francisco: Jossey-Bass.

Miles, M. B., & Huberman, A. M. (1994). *An expanded source book: Qualitative data analysis* (2nd ed.). Beverly Hills: Sage.

Perry, R., Lewis, C., & Akiba, M. (2002). *Lesson study in the San Mateo-Foster City School District.* Paper presented at the Annual Meeting of American Educational Research Association, New Orleans, LA. Retrieved July 15, 2002 from http://www.lessonresearch.net/articlescenter.

Portes, A. (1998). Social capital: Its origins and applications in modern sociology. *Annual Review of Sociology, 24*, 1–24.

Putnam, R. (2000). *Bowling alone: The collapse and revival of American community*. New York: Simon and Schuster.

Southwell, B., & White, A. (2004). Lesson study professional development for mathematics teachers. In M. Hoines & A. Fuglestad (Eds.), *Proceeding of the 28th conference of the International Group for the Psychology of Mathematics Education* (Vol. 1, p. 355). Bergen: Bergen University College.

Stein, M. K., Smith, M. S., Henningsen, M. A., & Silver, E. A. (2000). *Implementing standards-based mathematics instruction: A casebook for professional development*. New York: Teachers College Press.

Wiburg, K., & Brown, S. (2007). *Lesson study communities*. Thousand Oaks: Corwin Press.

Examining Change in Teacher Mathematical Knowledge Through Lesson Study

Diane Hobenshield Tepylo and Joan Moss

Despite evidence that traditional mathematics instruction—teachers providing step-by-step procedures followed by student practice—is not working for a large proportion of our students (Hiebert 1999), math teaching still overwhelmingly follows this traditional pattern (Fuson et al. 2005; Hiebert 1999). Many adults fear and lack competence in elementary mathematics (Tobias 1993). For example, Lamon (2006) estimated that 90% of adults are weak in proportional reasoning. Given that competence in mathematics skill is a requirement for many well-paying careers and a key to effective participation in the community (Steen 1997), improving mathematics instruction becomes an issue of social justice as well as of economic improvement (Moses and Cobb 2002).

There is growing evidence that changing instruction does, in fact, impact student learning; students benefit from instruction that adapts to students' understandings, that builds connections among mathematical ideas, and that fosters students reasoning for multiple solutions or strategies. Students receiving this kind of instruction have demonstrated better problem-solving skills than students taught in more traditional ways (Ross et al. 2002). However, the barriers to changing and improving mathematics teaching practices are well known and have proven to be difficult to overcome (Ross et al. 2002) and include teacher beliefs about mathematics and learning (Battista 1994), teacher isolation (Wallace and Louden 1994), and pressure to conform to existing school practices (Stein and Brown 1997). First and foremost amongst the barriers is the absence of the kinds of mathematics content knowledge that teachers require for this enhanced teaching (Ball et al. 2004; Sowder et al. 1998).

Lesson study, credited with successful mathematics teaching in Japan, is becoming increasingly popular in North America as a professional development method to address these challenges of change. For example, the National Staff Development Council identified lesson study as a potentially powerful approach to foster teacher change (Easton 2004). Indeed, there is a growing body of research that has pointed to the potential of lesson study to promote teacher change (Lewis et al. 2009). However, to date, there is little research on the mechanisms by which lesson

D. H. Tepylo (✉)
University of Toronto, Toronto, Canada
e-mail: diane.tepylo@utoronto.ca

L. C. Hart et al. (eds.), *Lesson Study Research and Practice in Mathematics Education*,
DOI 10.1007/978-90-481-9941-9_5, © Springer Science+Business Media B.V. 2011

study actually leads to improved teacher practices (Lewis et al. 2006). There is a continuing need to understand the specific mechanisms by which lesson study can promote changes in teachers' math content knowledge.

Lewis et al. (2009) present case study evidence that lesson study as a gestalt builds teacher knowledge. The present study provides a more fine-grained account of teacher learning, using a microgenetic approach—frequent observations with intensive phase-by-phase analysis (Siegler and Crowley 1991). The aim is to identify teachers' learning trajectories as they participated in lesson study, identifying changes in teacher knowledge in each of the phases of the lesson study cycle: goal investigation, planning, live research lesson, debriefing (see the chapter by A. Murata, this volume; Lewis et al. 2009).

Anticipated Teacher Learning in Each Phase of the Lesson Study Process

The Learning Mathematics for Teaching Project (Ball et al. 2008; Hill et al. 2008) identifies four important components of teacher knowledge: Common Content Knowledge (CCK), Specialized Content Knowledge (SCK), Knowledge of Content and Students (KCS), and Knowledge of Content and Teaching (KCT). Given the Japanese lesson study's focus on student learning (Lewis 2002), we hypothesized that KCS would increase throughout each phase of the lesson study. First we expected to see changes or growth in KCS during the *Investigating Goals* phase as teachers collaboratively consider goals for their students. In the second phase, *Planning*, we anticipated that, as the teachers experienced the detailed and exacting kind of planning that by necessity precedes a public research lesson, they would engage in the thinking that would support the development of KCS. In the third phase of lesson study, the *Research Lesson*, the teachers' role is to observe students and to collect data on student learning/behaviors. Thus, we anticipated that this would be a time where teachers would focus on student conceptions. Finally, our prediction was that the *Debrief/Reflection* phase would provide a powerful context for further growth in KCS as the teachers are given the opportunity to reflect on their observations of student thinking and to consider how the students' learning occurred as the result of instruction.

As the teachers gained new insights into student understandings, we anticipated that there would be potential changes in the teachers' Knowledge of Content and Teaching. We also predicted that there would be the potential for teachers to gain new content knowledge as they participated in the lesson study, particularly as they explored possible mathematical topics for the focus of the lessons. For this study we were not concerned whether this content knowledge was common to knowledgeable adults (CCK) or specific to teachers (SCK), so we chose to combine CCK and SCK into one type of knowledge, Teacher Content Knowledge (TCK), to clarify our analysis of multiple cycles of lesson study.

Finally, we predicted that with repeated cycles of lesson study, each domain of knowledge for teaching mathematics would be strengthened. Each kind of learning would be enhanced by repetitions, each building on learning from previous phases. In particular, teachers would become increasingly adept at identifying important student thinking.

Methods and Results

Our analyses for this research occurred as part of a broader project which involved a detailed microanalysis that tracked overall change in teaching practices of four teachers who participated in lesson study in the area of fractions (Tepylo 2008). For the present study we specifically analyzed changes in these teachers' MKT in the domain of fractions in order to test our hypotheses concerning the kinds of learning that would be supported at each phase of the lesson study cycle.

Teachers

This study examines the MKT learning of four grade 5 and 6 teachers in a small-town elementary school as they participated in three cycles of lesson study on teaching fractions. The school, in a rural school board, is a dual-track school (English and French immersion) resulting in classes in the English stream with 25–48% of students working on Individualized Educational Plans as compared to the 18% provincial average (EQAO 2009). The four teachers, Jeri, Brenda, Leslie, and Francis, ranged in age from 23 to 49 with three months to eight years of teaching experience. Before this study, the teachers had little exposure to mathematics professional development and none had any experience with lesson study.

Prior to the lesson study, the teachers identified fractions as the area they wished to explore during their lesson study. The first author then created a pre-test to probe teachers' understandings of fractions. The pre-test items did not include traditional calculations but entailed teachers talking aloud as they solved questions adapted from Lamon (2006) and Sowder et al. (1998) which examined how the teachers determined fractions in atypical contexts (see Fig. 1), how they compared fractions, and as well how they determined equivalent fractions.

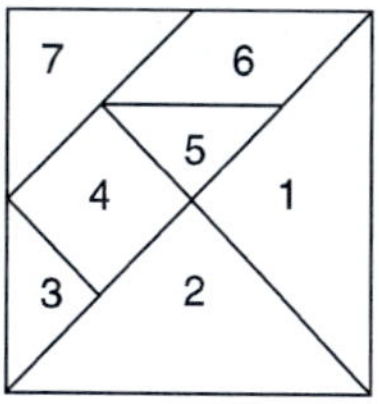

Fig. 1 The small square (#4) is what fractional part of the larger square?

Analysis of transcripts from this initial assessment revealed that the teachers were able to solve many of the questions procedurally but struggled more with the open-ended questions which could not be immediately solved with familiar procedures.

The first author also interviewed the teachers to investigate their knowledge of students' conceptions and their strategies for teaching fractions. Analysis of transcripts revealed that the teachers had little insight into students' conceptions of fractions. When asked how students might solve a problem, the teachers responded with who in the class could solve the problem, rather than how. Also, when presented with an item that showed examples of possible student strategies (see Fig. 3), the teachers dismissed the students' explanations of their thinking, arguing that the students were wrong, rather than attempting to determine how the students thought about the problem. The teachers mentioned that they were aware that they should use student strategies in mathematics instruction, but that they rarely did so. For example, in her pre-interview Brenda indicated that she disliked leading mathematical discussions:

> I'm the type of person that I just liked to do math. I don't want to explain, I don't want to talk, so I'm awful…. Every time I'm telling them [to explain their thinking], I feel like a hypocrite, because I know I hate it.

Data Collection and Analysis

With our interest in connecting changes in the teachers' learning to the different phases of the lesson study cycle, we videotaped all of the lesson study sessions and we analyzed more than 25 hours of session transcripts to track group changes as well as changes in the individual teachers (see Tepylo 2008). In addition, each teacher was interviewed at the end of the lesson study to assess changes in their understanding of fractions, student conceptions, and effective teaching strategies.

Within a two-month period, the group met formally eleven times, with many more informal meetings in the hall, the staffroom, and during a shared half-hour commute. Four of the formal meetings occurred during school hours, with coverage provided by administration or other staff. The remaining lesson study sessions occurred at lunch or after school. After an initial meeting to set the goals for the overall lesson study, the teachers conducted three cycles of lesson study designing and implementing three separate redesigns of an equivalent fraction lesson. In the sections that follow we describe the lesson study as it unfolded at Eastern Elementary and compare observed learning with our predictions.

Lesson Study at Eastern Elementary

Introduction to Lesson Study

At the first meeting, the first author, who acted as the facilitator for the study, described the lesson study process and provided all of the teachers with copies of

Teaching fractions and ratios for understanding (Lamon 2006), and materials from the Rational Number Project (Cramer et al. 1997). The group also watched a short video from the Mills College Lesson Study Group (Lewis 2005) demonstrating the highlights of the lesson study process.

Investigating Goals

At the first official lesson study meeting, the group discussed characteristics of their students and expressed their hope of increasing student engagement in mathematics. The teachers attempted random problems found in the Lamon (2006) text as a way of exploring possible goals for their study. In particular the group focused on problems that dealt with fractional parts of complex shapes, identifying the unit, fair sharing, and the quotient interpretation of fractions (each person gets m/n when n people share m things). After exploring the general context of fractions, the teachers finally decided on the topic of equivalent fractions for the focus of their lesson study. They noted that this was a topic that their students struggled with year after year.

Changes in Teacher Knowledge from Investigating Goals

Prior to this study, we anticipated that the participating teachers would demonstrate considerable increases in their Knowledge of Content and Students from collaboratively investigating goals for their students. However, the teachers did not discuss specific goals for student thinking in this phase, but concentrated on investigating novel mathematical contexts. The teachers initially struggled with the open-ended tasks in the Lamon text which could not be solved with a known procedure. However, as revealed in the protocols below, by persevering and by drawing on each other's knowledge of fractions, the teachers gained new insights into these problems.

The teachers initially grappled with Lamon's "reasoning up and down problems," determining the unit for a specified fraction and drawing. For example, in the problem illustrated in Fig. 2, the learner is challenged to find the whole if the

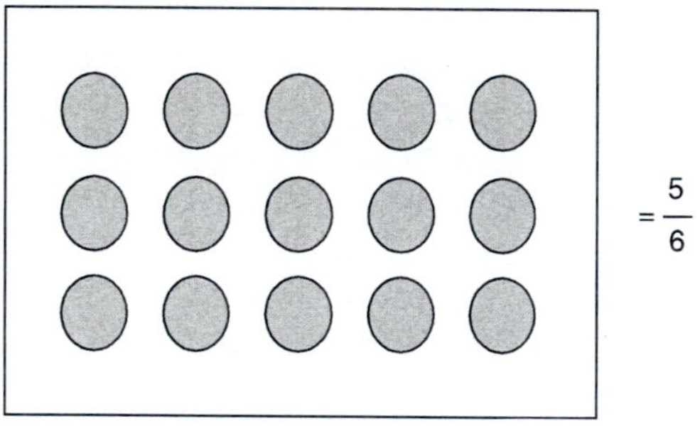

Fig. 2 Reasoning up: What is the whole?

figure represents 5/6 of the whole. Francis, who repeatedly requested "the procedure," eventually understood these types of questions by playing with the problem and scanning the text. She then assisted the other teachers:

Francis: Here is what you have to do. Use reasoning. You need an extra row of three. Five is the numerator. That was throwing me off.

Brenda: How do you know it's five sixths? How is it five sixths if that sixth row isn't there?

From their struggles with these types of problems during this phase, the teachers were able to solve this type of question on the post-test, even in novel situations such as determining the original angle if 3/4 of the arc is 135 degrees (Sowder et al. 1998, p. 204).

As the teachers examined fair-sharing problems in *Teaching fractions and ratios for understanding* (Lamon 2006), they initially questioned the value of this type of questions for students in grades five and six. The first author suggested that the questions could help students to develop an understanding of fractions as quotients: if m things are shared by n people, each person receives m/n of the things. The teachers did not initially understand the facilitator's explanation; it was only after the teachers explored a number of fair-sharing examples by drawing that they began to understand the concept:

Leslie: I see it now but I would have never thought to tell my kids.

Brenda: I wouldn't have figured that out.

Leslie: I wouldn't have figured that out. I wouldn't have connected the number of things and the number of people with the fraction.

This new Teacher Content Knowledge prompted the teachers to reconsider their teaching practices. Because the teachers themselves struggled to understand the quotient interpretation of fractions, they did not think that they could simply explain it to their students. Instead they needed to help their students figure it out.

Leslie: A word problem.... Should they draw it?

Brenda: Is that what you just did over there? [Looking at a drawing for the problem seven people sharing two pizzas.] I can see someone cutting this pizza into 7 pieces and this one into seven. You get one and one, two sevenths. That would give them the fraction.

Leslie: But it would be less efficient. We want them to be able to figure it out without drawing.

Brenda: They would actually have to draw first and then discuss the connections [...]

Francis: The discussion is the big idea, whereas here we don't sit down with the class and hold talk, we just give them more work.

The group concluded that students needed to attempt several fair-sharing questions, followed by a discussion to build this important generalization. This contrasts to the teacher's pre-interview where the teachers' indicated that they did not see much value in mathematical discussions.

In contrast to the rich learning of TCK and KCT, analysis of transcripts suggested only one possible instance of teacher learning about student understandings (KCS). When the facilitator specifically pointed out student understandings in the Lamon text, the teachers initially rejected a student's (the fictional Adam)

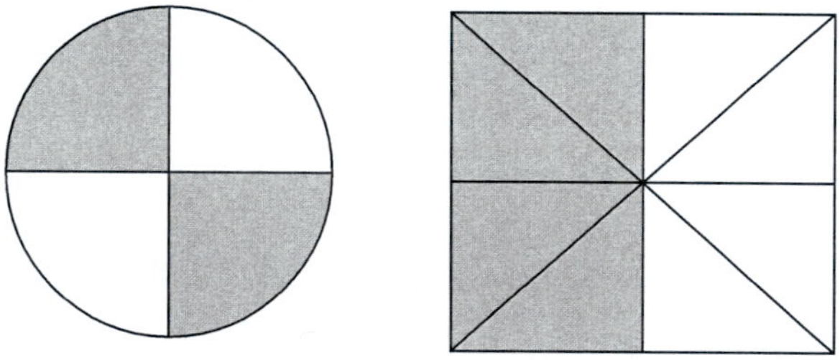

Fig. 3 Do the fractions name the same amount?

explanation regarding the question in Fig. 3 (Lamon 2006, p. 124), where the student argued that the fractions were the same (the shaded sections represented half of each of the two figures), but that the fractions also represented different quantities. At first, the teachers agreed with the first part of Adam's argument, but not the second. With prompting from the facilitator, the teachers came to realize that Adam's interpretation was possible, that same fractions do not always represent the same quantity. However, their responses suggest a greater focus on the mathematics than how students understand the concepts:

Francis: It could be half of a million or half of five. Which would you take? A half of something may be a different amount.

Brenda: Or a different shape.

Planning the First Research Lesson: Christmas Cookies

When planning their first research lesson, the teachers chose a lesson from their textbook. They modified a question asking students to identify the shaded fractional part of an array to a Christmas theme involving red and green cookies. The lesson was to start with four red and eight green magnetic "cookies" on the board. Students would be asked to identify the fraction of the cookies that were red and, by grouping the cookies into packages, determine equivalent fractions for the same amount of cookies. Then, in small groups, the students would be asked to group all the cookies (six red and eighteen green counters) into packages to determine equivalent fractions. The teachers anticipated that the manipulatives would assist the students to make sense of the problem. When prompted, the teachers identified one possible student interpretation of the question—students might deal only with whole numbers: i.e., one row or four cookies were red.

Teacher Learning During First Cycle Planning

Without the careful planning characteristic of Japanese teachers (Stigler and Stevenson 2001), little learning was evident during this first planning phase. Planning the lesson on their own, the teachers created a superficial lesson more theme-driven than mathematically driven. The teachers demonstrated little sense of the lesson within a projected learning trajectory or potential student outcomes. The only suggestion of increased knowledge during this phase resulted when the newest teacher, Jeri, learned about arrays from the other teachers.

The First Research Lesson: Christmas Cookies

Francis, Jeri, and Brenda observed this first lesson in Leslie's class, a class where about half the students had been identified with special needs. As the teacher created and rearranged arrays on the board, most students participated and correctly identified the fractions of the array. However, the small group portion of the lesson was not as successful, exposing the students' many misconceptions. Students had difficulty with the meaning of standard fraction notation. For example some students wrote random numbers separated with a fraction line; other students wrote part-to-part ratios. None of these difficulties had been anticipated by the teachers during the planning stage. Instead of ending the lesson reviewing equivalent fractions, the teacher finished the class by discussing fraction notation.

Teacher Learning During First Research Lesson

Because of the research lesson's stated focus on student understandings, we predicted that teachers' major gain in this phase would be to increase their Knowledge of Content and Students. However, the teachers' interpretations of the "Christmas Cookie" lesson revealed that they had not focused on student conceptions. Three of the teachers described which students could not do the problem, but offered no details of what the students were doing or thinking during the lesson. Only Brenda noted a student strategy.

> Brenda: I looked over [George's] shoulder. He was adding one to the top and four to the bottom [to determine equivalent fractions to 1/4]. I was thinking, "Where did that come from?" I would have never thought like that.

In contrast, Francis, Leslie, and Jeri focused primarily on teaching practices within the lesson, noting issues with content sequencing, manipulatives, and the layout of information on the blackboard (KCT).

Debrief/Reflection on the First Lesson: Christmas Cookies

In the debrief, most teachers glanced briefly at the students' written work and then listed the practical difficulties they had noted while in the classroom: the large size of the paper, the different shades of green counters, and the layout of information on the blackboard. Teachers discussed whether students were engaged in the lesson. Leslie, the classroom teacher, questioned whether these students could learn fractions because many even struggled with the comparatively simple two-digit addition and subtraction. She also questioned whether fractions were ratios; Leslie was adamant that fractions and ratios were completely different, and the students that produced part-to-part ratios were completely wrong.

Teacher Learning During First Debrief

We had predicted that in this phase the teachers would connect observed student learning with observed instructional strategies, thereby increasing KCS and KCT. Although the teachers seemed impressed with Brenda's insight on how George created equivalent fractions, they were more focused on their new insights about teaching, and did not make connections between teaching strategies and student learning. The fraction-ratio debate during the debrief suggested the possibility of new Teacher Content Knowledge. However, because this discussion ended without consensus, and because the issue never arose again, it is unclear whether new TCK developed during the first debrief.

Planning the Second Lesson: Fair-Sharing Pizza

Because the teachers were dissatisfied with their first lesson, they proceeded to plan a second lesson with the same goals as the first. The teachers revisited the Lamon resource, deciding to use a fair-sharing problem to address basic fraction concepts as well as equivalent fractions. The teachers planned a lesson that concentrated on different fraction labels for the same amount of pizza. Due to time constraints, two teachers planned the lesson, incorporating student names in the problems so as to engage the students. The teachers predicted that the students would initially draw pictures to understand equivalent fractions and might repeatedly add the numerators and denominators to create equivalent fractions as Brenda had observed in the first research lesson.

Teacher Learning During the Second Planning Session

As with the first planning session, the teachers quickly planned the lesson. They incorporated their knowledge from the first debrief: the importance of content sequencing and their one-student strategy for equivalent fractions. This inclusion may represent a deepening of these understandings during this phase; however, analysis of the transcripts provided only one clear example of new teacher learning. Searching the Lamon (2006) resource for a problem for the second lesson, Leslie discovered a common student difficulty with fraction multiplication: multiplying by a fraction creates a smaller answer in contrast to multiplying by a whole number which creates a larger answer. Francis was shocked by her new understanding of the mathematics and of how students might struggle, “I am 49 years old, and it never occurred to me before if you multiply something that it doesn’t always make it bigger!”

Second Research Lesson: Fair-Sharing Pizza

The second research lesson occurred in Francis’ grade 6 classrooms using joint preparation time to allow Jeri, Brenda, and Leslie to observe. The lesson started by asking students how much pizza each student would receive when two students shared a pizza fairly. Many students automatically knew the answer, but had difficulty justifying their answer. With teacher prompting, the students drew pictures and generated equivalent fractions by cutting the pizza into smaller pieces to justify their answers. Students then determined how much pizza each student would get if four and then eight students shared one pizza generating equivalent fractions for one-quarter and one-eighth respectively. During the lesson, one of the weakest students exclaimed, “I get this now!” However, the last question, added at the last minute to the lesson, posed the problem of sixteen students sharing two pizzas which confused students and teachers alike.

Teacher Learning During Second Research Lesson

The second research lesson provided more evidence of the predicted learning about students (KCS). Two teachers noted student strategies, documenting students’ successful use of drawings, the repeated addition strategy found in the first lesson as well as the traditional procedure for finding equivalent fractions. The teachers also noted how a weaker student, who was verbally able to name the fractions, seemed to struggle with the mathematical notation.

Second Lesson Debrief/Reflection: Fair-Sharing Pizza

The closing question of the second lesson, sixteen students sharing two pizzas, inadvertently raised confusion over the referent unit which became the focus of the debrief discussion: was the answer two-sixteenths or two thirty-seconds; was the unit both pizzas or one pizza? During the debrief, the teachers struggled with the mathematics and teaching implications of the question; they considered whether the wording of the question could have been clearer. Did “how much” automatically indicate that the answer should be a fraction? Should the question have been worded “of a pizza,” “of the pizzas,” or simply “How much pizza?”

Teacher Learning During Second Debrief

Analysis of the transcripts from the second debrief found that, in contrast to the first debrief, teachers now made the anticipated connections between student understandings and teaching: for fifteen minutes teacher alternately discussed teaching strategies and student conceptions. The teachers first noted that the drawings and the food context helped the weaker students understand equivalent fractions. The teachers enthusiastically shared the three student strategies they had observed emerging during the lesson. The teachers debated whether their planned starting place of one-half was a good choice. They wondered if the one-half context encouraged the students to operate automatically without reflecting. The teachers' reflected on how careful question wording might improve student learning and elicit a variety of student strategies. All this suggests the teachers' growth in KCT and KCS during the second debrief.

Our analysis of this phase also uncovered evidence of new Teacher Content Knowledge. The last question of the second live lesson (sixteen students equally sharing two pizzas) inadvertently introduced the very challenging idea in the conception of a unit in fractions. The teachers and the students were confused by the two feasible units. Should the number of pizzas eaten by each student be compared to the number of slices in both pizzas, or to the number in just one pizza? Analysis of the transcripts found that all four teachers initially struggled with the idea, but three of the teachers grew comfortable with both understandings.

Brenda:	Isn't there 32 pieces? So out of one pizza I get one sixteenth but out of two pizzas I get two thirty-seconds of the pizza. But two sixteenths of a pizza is....
Francis:	I would have two thirty-seconds of a pizza, [...] one from each pizza.
Brenda:	But if you were adding it together you would get two sixteenths.
Jeri and Leslie:	I see both.

In addition to deepening the teachers' understanding of the unit, this discussion may have increased the teachers' ability to determine the validity of mathematical arguments, an important component of TCK.

Planning the Third Lesson

After a three-week break for Christmas, the teachers began preparing for a third research lesson. Dissatisfied with each of the previous lessons, the teachers searched for a fresh instructional approach. They examined a variety of strategies from their own teaching experiences, the provided resources, and the school's small resource library before deciding to adapt an activity which involved shading equivalent areas on grids, each with a different number of divisions (Van de Walle and Folk 2005, p. 240). The teachers changed the grid activity to paper folding, hypothesizing that folding the same shaded area would highlight the unchanging area. The teachers hoped that the activity would support the students in building fluency to move between equivalent fractions and links between paper folding and the standard algorithm for determining equivalent fractions.

Teachers brainstormed possible student responses to the instructions and developed prompts to remedy possible student misconceptions. They also generated questions to focus student attention on the unchanging area. After a teacher-led exploration of equivalent fractions involving one-quarter, students were to work with a partner to explore equivalent fractions for three-quarters. Brenda tried this lesson in her grade 5–6 classroom and reported back that the problem engaged her students, but that it was important to stop with 32 squares; as she explained after five folds, her students' focus became the repeated folding rather than the mathematics.

Teacher Learning During the Third Planning Session

The planning of this third lesson approached the kind of detailed, well-considered processes associated in the literature with Japanese lesson study (Stigler and Stevenson 2001). Analysis of the third cycle planning transcripts revealed that, for the first time when planning a lesson, teachers built new Knowledge of Content and Students. They spontaneously suggested student strategies including the possibility that students might fold the paper in different directions and that students might need their attention directed to the unchanging shaded region. They even tried out their ideas in one of the classrooms to learn more about student responses and to help with the lesson planning.

The teachers also uncovered additional student thinking while investigating the possibility of using the folded paper to teach fraction addition in the third research lesson. Brenda and Leslie were surprised to learn that students often added both numerators and denominators:

Brenda: When adding, they don't know that the denominators need to be the same?

Leslie: They would add 2/5 + 1/2 and get 3/7s?

Francis: Exactly. And I don't know where the light went on in my head, but I think that's where kids break down in grade five and six, we spend so much time from [grade 1], teaching them to add on. They follow that rule and that rule doesn't change until they get to fractions, and then someone tells them they can add the top but not the bottom. It just throws them right off. Their whole rule of addition is now in question. If we went to another country and a red light suddenly meant go…. I am not driving in this country, I am going to get killed. I think it is the same with math […]

Brenda: It just occurs to me that people need to have a sense of what an equivalent fraction is before they can add fractions with unlike denominators.

In addition to these increases in Knowledge of Content and Students, the search for a more effective context for teaching equivalent fractions led to the concept of equivalent areas, an increase in Teacher Content Knowledge and Knowledge of Content and Teaching. The teachers had previously skirted the concept of equivalent areas while discussing the fractional parts of complex shapes during the initial goal-setting phase, but even with this early priming, thinking about equivalent fractions as equivalent areas was initially challenging:

> Leslie: I need to read this about five times.... What are we naming it here and here? ... It's not about the name; it's about the area!

The context of equivalent areas for equivalent fractions allowed the teachers to build previous KCT. The teachers built on a strategy they had previously used for introducing fractions, paper folding, and extended it to a new context, teaching equivalent fractions:

> Brenda: I like this. I never even thought of doing that [for students in grades 5 and 6]. I thought they had some background. I always do paper folding successfully in grade 4, but until I was sitting here folding, thinking "Why don't I do that in grade 5/6?"

The Third Research Lesson

Eager to share their learning with their colleagues, the teachers arranged a public lesson attended by eight visitors: two staff from their school, six teachers from other schools, and two special assignment teachers from the board office. As lesson study was new to these visitors, the afternoon began with an introduction to lesson study and the visitors were asked to note students' approaches to the problems.

Because the teachers were worried about student behavior with an audience, they selected Francis' class for the third lesson and Leslie was chosen to teach. The initial part of the public lesson went well, albeit more teacher-centered than the plan specified. Instead of an exploration, students were led through the paper folding. The students folded a paper in four and then shaded in one quarter. Almost all students were engaged, and the teacher's focus on the unchanging area seemed to help the weaker students understand how 1/4, 2/8, 4/16, and 8/32 were equivalent. Difficulties arose when students were asked to shade 3/4 of the folded paper, which now had 32 squares. Seeing some students struggle, Leslie demonstrated how to shade in two more corners to make 3/4. Even as students copied the shading, many students could not see 3/4 because of the numerous small squares. At this point, one of the stronger students provided the traditional procedure—multiplying the numerator and the denominator by the same number. The teacher then departed from the original lesson plan and focused on helping students to learn and use this strategy with no references to the paper folding. Only a handful of students participated in the remainder of the lesson.

Teacher Learning During the Third Research Lesson

During the third research lesson, the predicted learning about student thinking (KCS) was identified for three of the four teachers. The three teachers described specific examples of student thinking, including the difficulty that the students encountered in their attempts to identify 3/4 when the paper had multiple folds, and, more important, the misconception held by some of the students that the folding of the paper caused the area to change. Furthermore, Francis noted that three-quarters was a less friendly number for students than one-quarter, and Brenda noted how one student flipped over his paper muttering "The white is three-quarters. Why can't I just use that?" Interestingly, the visiting teachers, experiencing lesson study for the first time, provided no descriptions of student thinking despite specific instructions to do so prior to the lesson.

Third Debrief/Reflection

Following the research lesson, the group reconvened in the teacher preparation room to debrief, starting with the teacher of the lesson, then the other organizing teachers, followed by the visiting educators, and finally the discussant. The organizing teachers outlined their learning through the lesson study process and what they noticed about the lesson. The group discussed the reduction in student participation between the beginning and end of the lesson. The guests shared their perceptions of the lesson study and their favorite strategies for teaching fractions, with one math-phobic guest reflecting that lesson study was exactly what she needed to improve her understanding of mathematics. The discussant, a former mathematics special assignment teacher, provided suggestions for improving the lesson, including student-to-student discussion, which had been included in the jointly created plan but not implemented in the actual public lesson. The discussant challenged the group to explore how to help students create connections between the paper folding and the traditional algorithm for finding equivalent fractions.

Teacher Learning During the Third Debrief/Reflection

The prediction of deepening teachers' knowledge of students (KCS) and teaching (KCT) during the debrief was reaffirmed in the third cycle at Eastern Elementary. As the organizing teachers shared their observations of student learning, they often built on each others' descriptions and considered the implications of the student thinking for teaching. For example, when Leslie described one student who needed to draw lines on the folds to see and count the squares, Francis noted two other students that had done the same, remarking that teachers needed to be aware of student needs. Even as the group repeatedly commended the value of teaching equivalent fractions with equivalent areas, the teachers noted that some students required additional work with area before this lesson.

The reduction in student participation was initially dismissed by Leslie and Francis as an unavoidable fact of classrooms, "In every class, there are only four or five participants. It happens in every class." However, other participants connected the change in participation to the change in fractions, and to the abandoning of the paper. Slowly a consensus seemed to develop that starting the three-quarters portion of the lesson with an unfolded sheet of paper would have maintained the goal of more student participation.

One goal of this research lesson was to connect the paper folding to the standard procedure, but the lesson plan did not consider how to make those links. Not surprisingly, the connection did not come out in the public lesson. It was only in the debrief that Brenda seemed to make the connection, "Folding the paper in half is like multiplying the numerator by two and the denominator by two," because twice as many of pieces were now shaded out of twice as many pieces. The debrief's layering of multiple observations and numerous insights seemed significant in developing this important understanding.

Teacher Learning Through Lesson Study

The above microanalysis details the teachers' substantial teacher MKT learning throughout the fraction lesson study and is summarized in Table 1. Pre-post interviews corroborate these increases in knowledge. The teachers demonstrated

Table 1 Teacher learning throughout the lesson study

Cycle	Phase	Examples of teacher learning	
Investigating goals		TCK	Reasoning up and down, fractions of complex shapes, quotient interpretation, half is not a half
		KCT	Teaching quotient interpretation, sequencing decimals and fractions
Cycle 1: Christmas cookies	Planning 1	TCK	What is an array?
	Research lesson 1	KCS	Repeatedly adding numerator and denominator (Brenda)
		KCT	Content sequencing, issues with manipulatives, layout of information on blackboard
	Debrief 1	KCS	Repeatedly adding numerator and denominator (shared)
		KCT	Building on: Sequencing, manipulatives, layout of information on blackboard
		TCK	Fractions as ratios?
Cycle 2: Fair-sharing pizzas	Planning 2	TCK	Multiplication does not always make things bigger
	Research lesson 2	KCS	Drawings, repeated addition, traditional procedure (Brenda, Leslie)
	Debrief 2	KCS	Drawings, repeated addition, traditional procedure (all)
		KCT	Wait time, number choice, drawing, question wording
		TCK	What is the unit?
Cycle 3: Paper folding	Planning 3	KCS	Different folding strategies, adding denominators
		KCT	Paper folding for equivalent fractions, preparing prompts
		TCK	Equivalent fractions as equivalent areas
	Research Lesson 3	KCS	Drawing lines to see segments, difficulties with 3/4, area changed
	Debrief 3	KCS	Multiple ways to see fractions on paper
		KCS/T	Connecting student participation to paper folding
		KCT	Equivalent areas for teaching equivalent fractions, connecting paper folding to tradition procedure

important gains relating Knowledge of Content and Teaching to Knowledge of Content and Students. On the pre-assessment, the teachers identified no student strategies. In contrast, on the post-assessment, all four teachers identified some student conceptions and identified problematic fraction concepts. They described more strategies for teaching equivalent fractions and provided detailed examples of using them in their classrooms. Additionally, Leslie and Brenda provided rich details of using student thinking and discussions to build important mathematical ideas.

The teachers also demonstrated more Teacher Content Knowledge on the post-test. They answered more questions using a wider range of strategies and fraction contexts that included benchmarks, and the quotient interpretation of fractions. Additionally, when presented with novel mathematical justifications on the post-test, the teachers did not immediately reject them as they had done on the pre-test. Instead the teachers attempted to explain what the individual in the question was thinking. For example, when presented with a pizza-sharing question (Lamon 2006, p. 66) with student answers, Brenda immediately identified unit confusions. In addition, all the teachers expressed more confidence and less frustration on the post-test.

Reexamining the Mechanisms of Lesson Study

We began this study by anticipating how the phases of lesson study might support increases in teachers' mathematical knowledge for teaching. In this section we highlight particular processes that seem to generate the most productive changes in teacher knowledge. The first phase, *Investigation of Goals*, only occurred once at the beginning of our three cycles, but was very productive for increases in Teacher Content Knowledge. As the teachers investigated possible goals within the domain of fractions, they grappled with many new fraction contexts. In this early phase, even specific focus on student thinking by the facilitator resulted in more consideration of the mathematics (TCK) than of students (KCS).

The second phase, *Planning*, produced the anticipated rich learning of KCS only in the third cycle. The superficial planning in the first two cycles resulted in little evidence of teacher learning. Our teachers, like those investigated by Fernandez et al. (2003), seemed to lack the disposition and/or knowledge to plan a well-considered lesson built around student thinking. By the third lesson, however, our teachers were able to plan a detailed student-focused lesson; analysis of this phase provided numerous examples of teacher learning in KCS and also KCT and TCK. Similarly, our teachers initially had difficultly focusing on, and learning from, student thinking during the *research lesson*, but this increased through repeated cycles of lesson study. In the first research lesson, only one teacher identified a student strategy but after the public third lesson, three lesson study teachers independently described student conceptions.

The anticipated focus on student thinking during the debrief phase was found in each cycle in this fraction lesson study, but the amount and depth of learning seemed to increase with each cycle. For example, Brenda's sharing of a student strategy during the first debrief seemed to lead to more teachers observing student strategies. Only after three *repetitions* of lesson study were all four teachers able to provide examples of student conceptions in the post-test.

Interestingly, we observed learning of Teacher Content Knowledge during the debrief for the first and second lessons. Students and teachers struggling with the mathematics during these research lessons appeared to highlight the teachers' own mathematical issues which were then discussed during the debrief. One possible source of this teacher content learning may be the student conceptions themselves: as the teachers noticed student difficulties, they needed to reexamine their own mathematical understanding to understand the student thinking.

Discussion

Lesson study consists of a complex web of activities that have the potential to support professional development. Although limited, this study examining the building of four teachers' knowledge over two months suggests several important considerations of teacher learning through lesson study. Significantly, teachers new to lesson study may not necessarily focus on student conceptions, a finding consistent with the North American teachers studied by Fernandez et al. (2003). However, the teachers' initial focus on the mathematics deepened their Teacher Content Knowledge, which may have enabled the teachers' ability to see student conceptions in later cycles. It seemed that the teachers needed a richer mathematical understanding than the step-by-step procedures demonstrated in the pre-test in order to see students' alternative strategies.

Despite the teachers' gradual learning to see student understandings, the evidence from this small study suggests that the concentrated examination of student strategies and conceptions is an important mechanism for teacher learning. In our study, observing students using non-traditional procedures appeared to compel teachers to reconsider their own mathematical knowledge (Teacher Content Knowledge), and how mathematics could be taught (Knowledge of Content and Teaching). As teachers discovered the range of student strategies for solving fraction problems (KCS) their conceptions of effective teaching also seemed to expand (KCT). This suggests possible changes in beliefs. Although beliefs are not specifically analyzed in this study, these possible belief changes are important because beliefs have been identified as a significant barrier to teacher change (Battista 1994). These tentative observations will be investigated in further research.

Although longitudinal studies of larger groups of teachers are necessary to determine if the focus on mathematics and student thinking leads to long-term teacher change, this study suggests that the focus on student conceptions through lesson study can be a powerful supporter of teacher change. The central findings of this study have implications for teachers undertaking lesson study for the first time. Limitations of teacher mathematics knowledge are well-documented (Hill 2004; Ma 1999) and may affect teacher ability to understand student thinking. Additionally, while a focus on student thinking is an integral component of the lesson study process, teachers new to lesson study may not always be amenable to a careful focus on student understanding (Fernandez et al. 2003). In this study, learning to see and understand student conceptions was a gradual process and required repeated research lessons and debrief sessions to develop. This gradual growth to see and learn from student understandings reinforces the assertion of Lewis et al. (2006) that the summative evaluation of lesson study needs to await further studies into its essential mechanisms, and for teachers to develop greater abilities to analyze student understandings.

References

Ball, D. L., Bass, H., & Hill, H. C. (2004). Knowing and using mathematical knowledge in teaching: Learning what matters. Southern African Association of Mathematics, Science, and Technology Education. http://mscenters.org/mmla/fall06/knowing.doc. Accessed 12 July 2008.

Ball, D. L., Thames, M. H., & Phelps, G. (2008). Content knowledge for teaching what makes it special? *Journal of Teacher Education, 59*(5), 389–407.

Battista, M. T. (1994). Teacher beliefs and the reform movement in mathematics education. *Phi Delta Kappan, 2*, 463–470.

Cramer, K., Behr, M., Post, T., & Lesh, R. (1997). *Rational Number Project: Fraction lessons for the middle grades—level 1*. Dubuque: Kendall/Hunt Publishing Co.

Easton, L. B. (Ed.). (2004). *Powerful designs for professional learning*. Oxford: National Staff Development Council.

Fernandez, C., Cannon, J., & Chokshi, S. (2003). A US-Japan lesson study collaboration reveals critical lenses for examining practice. *Teaching and Teacher Education, 19*(2), 171–185.

Fuson, K. C., Kalchman, M., & Bransford, J. D. (2005). Mathematical understanding: An introduction. In M. S. Donovan & J. D. Bransford (Eds.), *How students learn: History, mathematics and science* (pp. 217–256). Washington: National Academic Press.

Hiebert, J. (1999). Relationship between research and the NCTM Standards. *Journal for Research in Mathematics Education, 31*, 3–19.

Hill, H. C. (2004). Professional development standards and practices in elementary mathematics. *Elementary School Journal, 104*(3), 215–231.

Hill, H. C., Ball, D. L., & Schilling, S. G. (2008). Unpacking "pedagogical content knowledge": Conceptualizing and measuring teachers' topic-specific knowledge. *Journal for Research in Mathematics Education, 39*(4), 372–400.

Lamon, S. J. (2006). *Teaching fractions and ratios for understanding: Essential content knowledge and instructional strategies for teachers*. (2nd ed.). Mahwah: Erlbaum.

Lewis, C. (2002). Does lesson study have a future in the United States? *Nagoya Journal of Education and Human Development, 1*(1), 1–23.

Lewis, C. (2005). *How many seats: Excerpts from a lesson study cycle*. Oakland: Mills College Study Group.

Lewis, C., Perry, R., & Murata, A. (2006). How should research contribute to instructional improvement? The case of lesson study. *Educational Researcher, 35*(3), 3–14.

Lewis, C., Perry, R., & Hurd, J. (2009). Improving mathematics instruction through lesson study: A theoretical model and North American case. *Journal of Mathematics Teacher Education, 12*(4), 285–304.

Ma, L. (1999). *Knowing and teaching elementary mathematics*. Mahwah: Erlbaum.

Moses, R., & Cobb, C. (2002). *Radical equations: Civil rights from Mississippi to the Algebra Project*. Boston: Beacon Press.

Ross, J. A., McDougall, D., & Hogaboam-Gray, A. (2002). Research on reform in mathematics, 1993–2000. *The Alberta Journal of Educational Research, 48*(2), 122–138.

Siegler, R., & Crowley, K. (1991). The microgenetic method: A direct means for studying cognitive development. *American Psychologist, 46*, 606–620.

Sowder, J. T., Phillip, R. A., Armstrong, B. E., & Schapelle, B. P. (1998). *Middle-grade teachers' mathematics knowledge and its relationship to instruction*. Albany: SUNY Press.

Steen, L. A. (1997). *Why numbers count: Quantitative literacy for tomorrow's America*. New York: The College Board.

Stein, M. K., & Brown, C. (1997). Teacher learning in a social context: Integrating collaborative and institutional processes with the study of teacher change. In E. Fennema & B. S. Scott (Eds.), *Mathematics teachers in transition* (pp. 155–192). Mahwah: Lawrence Erlbaum.

Stigler, J. S., & Stevenson, H. S. (2001). How Asian teachers polish each lesson to perfection. In M. Gauvain & M. Cole (Eds.), *Readings on the development of children* (pp. 181–194). New York: Worth Publishers.

Tepylo, D. R. H. (2008). *Investigation the effects of lesson study*. University of Toronto, Toronto.

Tobias, S. (1993). *Overcoming math anxiety*. New York: W. W. Norton & Company.

Van de Walle, J. A., & Folk, S. (2005). *Elementary and middle school mathematics: Teaching developmentally, Canadian edition*. Toronto: Pearson.

Wallace, J., & Louden, W. (1994). Collaboration and the growth of teacher's knowledge. *Qualitative Studies in Education, 7*(4), 323–334.

Response to Part I: Jumping into Lesson Study—Inservice Mathematics Teacher Education

Akihiko Takahashi

These chapters describe how four groups of US teachers and researchers attempted to replicate Japanese lesson study within their school systems and what they learned through the process. Although small, notable changes in mathematics teaching and learning are reported in each chapter, and they seem promising. For example, Tepylo and Moss report that the teachers in their study gained mathematical knowledge by going through lesson study. Olson, White, and Sparrow argue that the process of lesson study encourages teachers to reflect on their own classroom practices.

Although all the authors conclude that lesson study has the potential to provide teachers with opportunities to learn to really *see* student learning and to critically reflect on their own teaching practices, these impacts may vary depending on the support that the teachers receive during the lesson study cycle. Meyer and Wilkerson argue that there are potential factors in supporting teachers to increase their knowledge for teaching from the process of lesson study such as anticipating students' responses while planning lessons. Japanese educators confirm that anticipating student responses, which includes not only several correct answers but also typical misunderstandings or wrong answers to the question, is one of the crucial processes while planning lessons (Takahashi et al. 2005; Watanabe et al. 2008). Like any other intervention, the impact of lesson study will vary depending upon the quality of the support given *during* the process, and greatly contributes to the quality of teacher learning.

In addition to the quality of the support, the authors argue that group dynamics among teachers in a lesson study group can affect the impact of the lesson study. In fact, one group in their study did not complete a lesson study cycle due to ineffective group dynamics among the teachers.

In a similar vein, Hart and Carriere argue that the role of knowledgeable others/facilitators is crucial for making lesson study meaningful and powerful. Their findings confirm that lesson study contributes to improved teaching and learning if teachers receive adequate support during the lesson study process. Although lesson

A. Takahashi (✉)
DePaul University, Chicago, USA
e-mail: atakahas@depaul.edu

L. C. Hart et al. (eds.), *Lesson Study Research and Practice in Mathematics Education*,
DOI 10.1007/978-90-481-9941-9_6,

study has been defined as a participant-centered and teacher-led form of professional development (Fernandez and Yoshida 2004; Lewis 2000, 2002; Murata and Takahashi 2002; Takahashi and Yoshida 2004), simply giving time for the teachers to follow the process of lesson study is not sufficient to expect substantial improvement of teaching and learning. Moreover, the visible impacts of lesson study may not be detectable until the teachers experience multiple lesson study cycles, even when they are given adequate support from their facilitators and knowledgeable others.

While various reports suggest that lesson study is an ideal way to improve teaching and learning as it allows teachers to work collaboratively to seek better classroom practices for their students and it allows them to find ways to make a successful lesson happen (Stigler and Hiebert 1999; Stigler and Hiebert 2009; Wei et al. 2009), a question remains as to why lesson study requires such a long-term commitment from teachers and schools in the USA, as well as why strong support from an outside expert is needed in order to see the impacts on both teaching and learning.

My reaction to this question is two-fold. There is a lack of experienced lesson study practitioners outside Japan, and it is difficult to address the two issues of (1) implementing a new teaching approach while (2) undertaking a new form of professional development simultaneously.

First, the lack of experienced lesson study practitioners is a challenge that Japanese teachers usually do not encounter, since lesson study is the major form of professional development in Japan. Most practicing teachers in Japan have experience in developing lesson plans for research lessons and have the skill to write lesson plans for a lesson study, which is different in both format and contents from lesson plans that Japanese teachers use in everyday teaching. In fact, Japanese teachers use the term *Ryakuan*, which means *simple version of the formal lesson plan*, to distinguish them. How do Japanese teachers learn to prepare such detailed lesson plans for lesson study? Moreover, how do they learn to conduct a lesson study? Rather than reading books and watching videos, Japanese teachers learn lesson study by experiencing public research lessons, observing lessons, and participating in post-lesson discussions. Japanese teachers usually do not begin lesson study by planning lessons by themselves.

During preservice years, Japanese teachers are introduced to the basics of lesson study, e.g., observing actual teaching with their peers and experiencing post-lesson discussions lead by professors and the cooperating teachers. Novice teachers are encouraged to first participate in public research lessons in order to establish the basic skills needed to observe research lessons and to provide constructive criticism in post-lesson discussions. They are expected to acquire these skills implicitly by watching experienced practitioners' behaviors during the research lessons and the comments given during the post-lesson discussion. The final comments given by the outside experts and instructional leaders at the end of the post-lesson discussion are considered to be one of the most valuable learning opportunities, not only for the novice teachers but also for all the research lesson participants.

Novice teachers are also encouraged to join a lesson-planning team for their school based lesson study in order to support the team in planning and in conducting the school based lesson study. After a while, Japanese novice teachers experience their first research lesson in their classroom with support from experienced teachers in the school building.

Due to the lack of an existing lesson study culture and experienced lesson study practitioners in the USA, beginning a lesson study is extremely challenging. In order to overcome these existing challenges, it is crucial for educational leaders and experienced practitioners to become able to answer critical questions similar to those that Hart and Carriere raised by gaining first-hand experience before asking other teachers to go through the lesson study process.

Second, it is also important to realize that through lesson study, US teachers and schools are being introduced not only to a new professional development approach, but also to a new way of teaching and often to different curriculum materials. This is similar to asking teachers to solve a new problem with a tool that they have never seen before. Teachers new to lesson study are often confused about what lesson study is. They merely see it as a new way of teaching or a new approach. Educators who organize workshops for introducing lesson study need to be knowledgeable about lesson study and how it can be used to improve teaching and learning.

In spite of these challenges, all four groups of authors reported evidence of some teacher growth which is often not observable in typical professional development in which a teacher merely receives information from the outside expert. Considering all four cases having implemented this very different professional development approach within the existing challenges teachers are facing in the schools, the findings are very promising. As long as previous efforts and experiences (such as those shared in these chapters) are considered, I believe the second generation of lesson study will have greater potential to impact teaching and learning.

Finally, the education leaders and researchers who support teachers through the lesson study process should keep in mind that the *idea* of lesson study is simple, but, as shown in these chapters, lesson study is not an easy jump for those teachers who have never experienced it before. Providing opportunities for observing research lessons and experiencing the post-lesson discussion with experienced practitioners is a desirable and often necessary first step.

References

Fernandez, C., & Yoshida, M. (2004). *Lesson study: A Japanese approach to improving mathematics teaching and learning*. New York: Routledge.

Lewis, C. (2000, April). *Lesson study: The core of Japanese professional development*. Paper presented at the annual meeting of the American Educational Research Association, New Orleans, LA.

Lewis, C. (2002). *Lesson study: A handbook of teacher-led instructional change*. Philadelphia: Research for Better Schools.

Murata, A., & Takahashi, A. (2002). Vehicle to connect theory, research, and practice: How teacher thinking changes in district-level lesson study in Japan. *Proceedings of the twenty-fourth annual meeting of North American Chapter of the International Group of the Psychology of Mathematics Education*, Athens, GA, USA, 1879–1888.

Stigler, J., & Hiebert, J. (1999). *The teaching gap: Best ideas from the world's teachers for improving education in the classroom*. New York: Free Press.

Stigler, J. W., & Hiebert, J. (2009). Closing the teaching gap. *Phi Delta Kappan, 91*(3), 32–37.

Takahashi, A., & Yoshida, M. (2004). How can we start lesson study? Ideas for establishing lesson study communities. *Teaching Children Mathematics, 10*(9), 436–443.

Takahashi, A., Watanabe, T., Yoshida, M., & Wang-Iverson, P. (2005). Improving content and pedagogical knowledge through *kyozaikenkyu*. In P. Wang-Iverson & M. Yoshida (Eds.), *Building our understanding of lesson study*. Philadelphia: Research for Better Schools.

Watanabe, T., Takahashi, A., & Yoshida, M. (2008). Kyozaikenkyu: A critical step for conducting effective lesson study and beyond. In F. Arbaugh & P. M. Taylor (Eds.), *Inquiry into Mathematics Teacher Education, Association of Mathematics Teacher Educators (AMTE) Monograph Series* (Vol. 5). San Diego: AMTE.

Wei, R. C., Darling-Hammond, L., Andree, A., Richardson, N., & Orphanos, S. (2009). Professional learning in the learning profession: A status report on teacher development in the U.S. and abroad. Technical Report. Washington: National Staff Development Council.

Part II
Emerging Issues from Lesson Study Approaches in Prospective Mathematics Teacher Education

Investigating Approaches to Lesson Study in Prospective Mathematics Teacher Education

Maria Lorelei Fernandez and Joseph Zilliox

As mathematics teacher educators, we are interested in creating and identifying contexts and teaching approaches that provide prospective teachers with experiences to help them develop rich understandings of reformed mathematics teaching. Our interests are aligned with recent calls to develop and study pedagogical approaches and experiences in teacher education (Graeber 1999; Grossman 2005). What we seek for prospective teachers who enroll in our mathematics education courses is what we want them to seek for their students. If mathematics teachers are expected to motivate their students' learning by arousing their curiosity, challenging their thinking, and engaging them actively in learning (Australian Association of Mathematics Teachers [AAMT] 2002; National Council of Teachers of Mathematics [NCTM] 2000), then we expect at least the same of ourselves, as teacher educators, in our work with future teachers of mathematics. During methods courses, prospective teachers may be presented with theoretical and research-based ideas; however, these novices often feel the discussions of theory and research are not sufficiently oriented toward practice and the methods courses are not intellectually substantive (Bransford et al. 2000).

One way we sought to intellectually challenge the prospective teachers enrolling in our mathematics methods courses was to design instructional experiences that build on the idea of 'lesson study,' a professional development practice that is highly valued among Japanese teachers (Stigler and Heibert 1999; Fernandez and Chokshi 2002; Lewis et al. 2006). We drew on central features of lesson study to design learning experiences involving what we think of as 'lesson study approaches' for teaching prospective mathematics teachers. These approaches may support prospective teachers' trial and analyses of their emerging understandings of theory and research-based instructional practices. Investigations on adaptations of lesson study across different contexts, including prospective teacher education, are needed (Lewis et al. 2006).

Herein we discuss our investigation of lesson study approaches for teaching prospective teachers at two different universities during semesters when they enrolled

M. L. Fernandez (✉)
Florida International University, Miami, USA
e-mail: mfernan@fiu.edu

L. C. Hart et al. (eds.), *Lesson Study Research and Practice in Mathematics Education*,
DOI 10.1007/978-90-481-9941-9_7,

in courses on learning to teach mathematics. One researcher worked with prospective secondary mathematics teachers in a microteaching setting arranged as part of an introductory mathematics education course. The other researcher worked with prospective elementary teachers in K-6 classrooms as part of their field experiences when enrolled in mathematics education methods courses. Our common purpose was to investigate elements of lesson study approaches in providing prospective teachers with opportunities for growth as teachers of mathematics.

Relevant Literature and Theoretical Perspective

Through use of lesson study approaches, we proposed to provide our teacher education students with authentic learning experiences to support their development as teachers of mathematics as part of their initial courses on learning to teach mathematics. According to Putnam and Borko (2000), an authentic experience for teacher learning is one that fosters the kinds of thinking and problem solving important to teaching. Four important aspects of teaching comprised in the proposed lesson study approaches included planning a lesson to meet particular mathematics goals and objectives, teaching the lesson to students lacking familiarity or understanding of the mathematics, reflecting on the lesson and student learning, and revising the lesson for future implementation. These aspects of teaching are routinely practiced by inservice teachers and are an important attribute of authentic instructional tasks in teacher education (Iverson et al. 2008). In addition, the lesson study approaches involved substantial cooperation among peers in completing these practices. Although the intended extent of cooperation may not be routinely practiced by inservice teachers, the meaningful discussion it was intended to foster has been found important in increased positive changes in practicing teachers' knowledge, skills, and classroom practices (Garet et al. 2001).

A theoretical perspective guiding this investigation of lesson study approaches has its basis in sociocultural theory (Vygotsky 1978). From this perspective, learning is a sociological process of participation in a group. Knowledge about teaching is thought to be socially constructed and students of teaching are thought to link theory with practice through jointly constructed learning communities (Frykholm 1998). Given our perceived centrality of group collaboration and learning within the lesson study approaches, a sociocultural perspective provides a valuable lens for investigating the design elements of these approaches. Of particular interest was Vygotsky's (1978) zone of proximal development (ZPD) that defines the difference between what an individual can do with help and what he or she can do without guidance. For Vygotsky, bridging the ZPD tended to require a teacher (or other similarly expert person) or a more capable peer.

Effective environments for human learning are thought to consist of four central features: learner-centeredness, knowledge-centeredness, assessment-centeredness, and community-centeredness (Bransford et al. 2000). Learner-centeredness refers to environments that attend to the learners' prior knowledge, skills, and beliefs while providing situations and opportunities for their continued thinking about, development, and adjustment of their ideas and skills. Knowledge-centeredness refers to environments that begin with concern for students' initial preconceptions and "focus on the kinds of information and activities that help students develop understanding of disciplines" (p. 136), including such activities as sense-making, metacognitive processing, and reflection. Assessment-centeredness refers to environments providing opportunities for feedback and revision in line with the learning goals. Community-centeredness refers to individuals learning from one another, continually striving to improve, and feeling connected to one another and the larger community. Our implementation of lesson study approaches with prospective teachers of mathematics was in part intended to align our teaching of these novices with recent conceptions of effective environments for human learning consisting of these four central features.

Design and Methods of Inquiry

The lesson study approaches under investigation were conducted with elementary and secondary prospective teachers enrolled in courses on learning to teach mathematics in their respective teacher education programs. In these courses, the prospective teachers were introduced to ideas, theories, and research relevant to recent reforms in teaching mathematics, materials and tools for teaching mathematics, and skills such as designing lesson plans. The lesson study experiences were intended to provide opportunities for the prospective teachers to put into practice and expand their understanding of what they were learning in their coursework. Table 1 delineates the design elements of the two lesson study approaches in the present investigation. We divided the design elements into five categories: lesson content, structure of experience, forms of feedback, implementation setting, and products.

For the secondary teachers, the lesson study approach involved teaching mathematics to their peers within the context of their initial course on learning to teach mathematics. The implemented approach is called Microteaching Lesson Study (MLS) (Fernandez 2005, 2010). MLS has been investigated with other groups of prospective teachers producing encouraging results for prospective teacher learning. These results include fostering changes in mathematics pedagogy and promoting learning through deliberation-in-process (Fernandez 2010) and supporting the development of technological pedagogical content knowledge among prospective mathematics and science teachers (Cavin 2007). The elementary teachers taught the school children they were working with during their weekly two-day field experiences. Teaching children in their classroom setting is typical of Japanese lesson study (Stigler and Hiebert 1999). Both the secondary and elementary prospective teachers were given information to guide their path through the phases of their lesson study experiences and outline the expectations for their group lesson study written and oral reports.

Table 1 Design elements of lesson study approaches

	Secondary MLS	Elementary LS
Lesson content	Overarching mathematics student learning goal, selected purposely by instructor for prospective teacher development Mathematics content for lessons selected by instructor so that prospective teachers lacked familiarity or knowledge Focus on student learning related to actual teaching	Overarching mathematics student learning goal, criteria proposed by methods course instructor Mathematics content for lesson agreed upon with classroom teacher in field experience Focus on student learning related to actual teaching
Structure of experience	Repeated cycles of planning, implementation, analysis, and revision involving use of videotaped lessons Collaboration within MLS group to include consultations with instructor as knowledgeable advisor	Repeated cycles of planning, implementation, analysis, revision Collaboration within lesson study group to include optional consultation with others
Forms of feedback	Feedback from instructor as knowledgeable advisor Feedback from peers in MLS group during teaching cycles Videotapes of implemented lessons No oral report in methods class if taught to student-peers[a]	Optional feedback from instructor or classroom teacher in cycles Feedback from lesson study group members during teaching cycles No lesson videotape Feedback from peers and instructors during oral report of work to their methods classes
Implementation setting	Reduced class size (5 to 10 students or student-peers)[a] sufficient for small cooperative group work (2 or 3 per group) Reduced lesson length (approx. 30 min)	Usual classes, in which completing weekly field experiences Usual lesson length (approx. 45 min)
Products	Written reflective report and lesson plans to be shared	Written reflective report Oral report of lesson study work to community of methods classes

[a] In the present study, the MLS involved student-peers

Secondary Prospective Teacher Lesson Study Approach

Thirty-six secondary prospective mathematics teachers participated in the investigation. These 36 teachers were enrolled, 18 each semester, in an initial course on learning to teach secondary school mathematics in an upper-division teacher education program. The lesson study approach, incorporated as part of the course, involved the novices in teaching mathematics to their peers within the course. Secondary school mathematics topics for the research lessons were selected purposefully, through the use of a survey, so that the peers being taught lacked familiarity, recall, and understanding of the mathematics. This produced a situation where the novice teachers also lacked knowledge of the content; a situation not uncommon to beginning secondary school mathematics teachers with whom we work.

The lesson study approach implemented with the secondary prospective teachers occurred over several weeks of each 15-week semester course. The teaching of the MLS lessons occurred during class time over three weeks to small groups of classmates from other MLS groups. Time for analysis and revisions to take place occurred from one teaching day to the next. Each semester, the instructor, one of the researchers, placed the 18 prospective teachers into heterogeneous MLS groups of three. MLS group assignments were based on participants' mathematics ability and ideas about teaching mathematics. Each MLS group was given a different mathematical relationship or concept to teach to their student-peers within approximately a 30-minute lesson. The overarching learning goal of the MLS mathematics lessons was to develop students' mathematical reasoning and ability to study patterns in constructing and justifying relationships or concepts.

Each semester the 18 prospective teachers were divided into three 'small classes' of six participants in such a way that only one member of each MLS group was in each of the three classes (see Fernandez 2005). The MLS lessons were taught within these small classes, in three different, nearby locations, so that all six MLS lessons were taught each week and thus could be concurrently revised and then retaught in a different small class, by another MLS group member, the following week. Group members taught and revised their lesson three times. After the initial research and planning of a group lesson, one MLS group member taught the lesson to their peers in their small class while being videotaped. Each member of the MLS group watched the videotape (individually or as a group), analyzed aspects of the lesson (individual reflections), and engaged in group discussions and revisions for the reteaching. Upon revision of the lesson, a second MLS member taught the lesson to the peers in their small class and the analysis and revision process was repeated. Finally, the lesson was taught by the third MLS member to a third group of peers and once again revised. The MLS groups produced written reports documenting the cycles of their MLS experience. Throughout the cycles, the instructor was available as a resource, observing lessons, watching videotapes, and providing feedback, primarily in the form of questions, for the groups to consider. The instructor engaged in postlesson discussions during the first or second cycles at least once with each group and posed select questions in writing during other cycles.

Elementary Prospective Teacher Lesson Study Approach

The elementary group consisted of 48 prospective teachers enrolled in either their second or third semester of a four semester field-based undergraduate program offering initial certification. For each of the first three semesters of the program, the prospective teachers completed two full days every week of supervised work in an elementary classroom and two days of university coursework. The fourth semester of the program consisted of traditional student teaching. Since the prospective teachers participating in this investigation were in their second or third semester of the program, they had varying degrees of experience in classrooms. Their lesson study took place during the school-based field experiences and was completed each semester in conjunction with their coursework.

Each elementary lesson study group consisted of three or four prospective teachers. The instructor, one of the researchers, created groups based on the grade level and location of the participants' field placements. Participants from different school locations were mixed purposely across the lesson study groups. As an overarching student learning goal, the research lessons were to involve hands-on activities that encouraged student discourse and group work in developing understandings of the mathematics being taught. The children were expected to do more than listening to the teacher. Each lesson study group was involved in three cycles of planning, teaching and observing, analyzing, and revising a mathematics research lesson. Individual lessons typically lasted between 45 minutes to one hour. Group members observed one another during their lessons and were expected to keep notes on the teaching and learning that occurred to be shared during the analysis phase of each cycle. All groups completed the three cycles within about three weeks.

Mathematics content topics varied from group to group. Lesson study groups selected the content to be taught in consultation with the regular classroom teachers. When possible during the lesson study experience, the university supervisor (one of the researchers) or one of the regular classroom teachers would contribute to the conversation about a research lesson. The lesson study groups submitted written reports of their work and presented oral reports of their experience at a gathering of all the participants. Each member of a group was expected to present some part of their group's oral report.

Data Collection and Analysis

The research design involved qualitative data collection and analysis. Data collection included the following: field notes of group planning, implementing, and analysis of the lessons; memos of informal conversations with group members; written reflective reports of the iterative cycles; videotapes of the lessons; videotapes of end-of-semester oral group presentations; and participant feedback surveys about the lesson study experiences. The authors initiated the data analysis by considering the data from the elementary and secondary school teachers separately. For each group of teachers, the analysis began with the coding of the prospective teachers' development, factors contributing to the development, and struggles within individual lesson study or MLS groups. The researchers used methodological and data source triangulation (Denzin 1984) to substantiate interpretations or clarify the developing meanings. At the next level of analysis, the researchers compared the findings arising from the secondary and elementary prospective teacher groups in relation to the four central features of effective environments for human learning discussed above.

Findings

In this section, we present the results of our investigation of the two lesson study approaches in relation to the four features thought to define effective environments for human learning. Our findings reveal the influence of design elements in each approach for facilitating the prospective teachers' learning about teaching mathematics.

Learner-Centeredness of the Lesson Study Approaches

For both approaches, the structure of the lesson study experiences and group products that guided individual learner and group contributions in thinking about and revising aspects of the research lessons were evidenced to contribute particularly to the learner-centeredness of the lesson study experiences. Within the MLS groups, the prospective secondary mathematics teachers were required to teach one lesson within the three MLS cycles and contribute individual reflections and group analysis and revision for each research lesson draft. Together their efforts supported the evolution of their research lessons and creation of their MLS written reflective reports. Through this process, the prospective teachers across all of the MLS groups were observed having opportunities to repeatedly think about, put into practice, and make adjustments to their prior and emerging knowledge, beliefs, and skills about teaching mathematics. For example, as part of their first lesson draft, a group teaching about ellipses began their lesson with a teacher demonstration of how to make an ellipse with two foci and a string of fixed length. After this introduction, the teacher gave the student-peers a couple of examples of ellipse equations with corresponding graphs to show the general ellipse formula. Through the analysis and revision cycles, including interactions with the course instructor, the teachers developed and trialed tasks that engaged their student-peers in explorations with physical models and multiple representations of ellipses on graphing calculators. The teachers were able to observe their student-peers engaging in constructing a general equation for ellipses through exploration of multiple representations of ellipses. This MLS group, as many others, commented on their learning to teach mathematics as follows:

> As a group we realized how important it was to make each lesson more student led rather than teacher led. Getting away from a teacher-centered classroom is hard, since that is how most of us learned. After seeing how a student-led lesson can improve a student's performance, we see how valuable it is in our teaching.

The iterative cycles provided opportunities for them to think about, trial and revise their ideas about teaching mathematics, helping them move beyond their prior knowledge developed through an apprenticeship-of-observation (Lortie 1975).

The elementary teacher lesson study approach was observed to provide similar opportunities for the prospective elementary school teachers. The sharing and negotiating that the lesson study groups participated in as part of the iterative cycles and creating their final reports were found to help them continue thinking about, putting into practice, and modifying their ideas about teaching mathematics. For example, a group teaching first-grade children to identify and classify shapes noted during analysis of the first lesson implementation that the children spent too much time listening to the teacher talk about the attributes of the shapes. Despite the teacher's explanations and modeling of attributes with the shapes, the children had difficulty sorting and classifying the blocks. In the revised lesson, the children were given the materials earlier in the lesson with less time given to the teacher show and tell. The prospective teachers noted that, as a result of the revisions, the children were more familiar with the characteristics of the blocks and had an easier time sorting the materials and creating their own patterns. The prospective elementary teachers as a whole recognized the merits of continued consideration and modification of their lessons for improved student mathematics learning. Observations reported by the prospective teachers included, "The changes to the lesson presentation that our group talked about proved to be valuable, and I saw how simple changes could make a big difference in a lesson." The prospective teachers also commented on how the collaborative iterative structure helped them make adjustments and build on one another's teaching experiences.

> I must say—it was terrific! It was organized, it was well planned, it was clear, and it made sense. [The third teacher] is a great teacher, but I cannot give him all the credit. Without [my] trail blazing and my experimentation gone bad, [the third] lesson could not have been as smooth as it was.[1]

Knowledge-Centeredness of the Lesson Study Approaches

The lesson-content design elements were found to play an important role in the knowledge-centeredness of both lesson study approaches. The overarching student learning goals for the elementary and secondary mathematics lessons and the particular mathematics content being taught provided direction for and influenced the sense-making and reflection about mathematics and the teaching of mathematics during the lesson study cycles. Within the MLS, the overarching student learning goal for the research lessons was to develop students' mathematical reasoning and ability to study patterns in constructing and justifying mathematical relationships or concepts, an area in which US teachers are weak (Jacobs et al. 2006). The mathematics content of the lessons involved mathematical ideas for which the prospective teachers lacked familiarity or understanding. The secondary prospective teachers were observed and reported as working diligently to learn and deepen their understanding of their mathematics topic in preparation for teaching it. One secondary teacher commented, "I tried to learn as much about the topic as possible prior to teaching the lesson so that I could answer any question given and explore ideas and connections more fully."

[1] Part of the elementary data was collected collaboratively with a science methods instructor (see Marble 2006).

Although the prospective teachers learned much about their mathematics topic while designing their first research lesson draft, questions raised as part of the lesson implementations often revealed to the teachers gaps in the depth of their understanding of the mathematics or in the strategies for eliciting from their student-peers mathematical reasoning for constructing and defending the mathematical concepts and relationships being taught. The following exemplifies our findings across MLS groups. When teaching about odd and even functions, one group's first lesson draft involved the teacher telling the students about the symmetry of graphs and then showing the students examples of odd and even functions to explain the properties that define each. Through the cycles of analysis and revision that followed, the MLS group considered whether they were engaging their students sufficiently in the reasoning process and in defining odd and even functions through their own observations. The teachers developed lessons that were more student-centered, allowing students to explore, analyze, discover, and explain properties of odd and even functions and create their own examples. In their written report, the group explained their learning:

> We focused on not giving away the properties of odd and even functions, letting the students discover these on their own through more group work. Having students create their own examples and share these with the class was also a great idea and went well.

The MLS was found to support the secondary prospective teachers' development of their understanding of the disciplines of mathematics and of mathematics teaching through their engagement in sense-making, metacognitive processing, and reflection about the mathematics and how to teach it in ways aligned with recent mathematics education reforms.

The lesson-content design elements of the prospective elementary school teachers' lesson study were different from those of the MLS implemented with the prospective secondary school teachers. The overarching learning goal for their research lessons was to involve hands-on activities that encouraged student discourse and group work in developing students' understanding of the mathematics being taught. The prospective teachers selected the mathematics content for the lessons, providing occasion for some groups to choose content with which they felt comfortable, as well as content for which they may have thought student learning would not be problematic. For these prospective teachers, their depth of sense-making and metacognitive processing of the mathematics or mathematics teaching was at times lessened. One representative example of our finding involved a group teaching about fractions by folding paper. When the children struggled to complete a task with fractions where they had to fold paper strips into various equal portions, the lesson study group revised the lesson by placing inked fold lines along the paper strips. The group claimed that the task became easier for the children, but they failed to recognize that it was now mostly a folding activity having lost much of the part–whole nature of the previous fraction task. Even though one of the group members was aware of this limitation, the dynamic of the group prevented this individual from exerting an altering influence. This group did not receive feedback during the iterative cycles from a knowledgeable advisor that could have supported giving serious consideration for the minority individual viewpoint. Had the prospective teachers felt challenged by the mathematics or their students' understanding of the mathematics, as was the case with the MLS groups, they may have engaged more thoughtfully in the sense-making about teaching mathematics and students' mathematics learning.

Assessment-Centeredness of the Lesson Study Approaches

As might be expected, the forms-of-feedback design elements were found to be of central importance in the assessment-centeredness of the lesson study approaches. Both approaches consisted of feedback from lesson study group members during the iterative cycles; however, other feedback elements differed considerably between each approach. As part of the MLS approach, the course instructor (a knowledgeable advisor) was part of the lesson analysis within the iterative cycles. Additionally, the research lessons were videotaped for use during the analysis phases of the cycles. For the prospective elementary school teachers, the lesson study approach did not require the involvement of a knowledgeable advisor during the iterative lesson study cycles and no lessons were videotaped. Instead, feedback from knowledgeable advisors (instructors) and peers from their elementary mathematics methods classes was provided during the presentation of their lesson study oral reports.

Triangulation of data sources revealed the positive influence of the knowledgeable advisor in the MLS groups' analyses and revision of their research lessons. MLS feedback surveys completed individually by the secondary prospective teachers made known their view that the instructor (knowledgeable advisor) feedback was important to their understanding and growth of their lessons. Many of the prospective secondary school teachers wrote comments such as, "Input from the instructor raised questions that helped us look at what to change. It helped us clarify how to 'discover-a-relationship' and was useful in incorporating graphing calculators to help enhance the lesson." These individual comments, triangulated with observation notes, written reports, and videos of the lessons, revealed the importance of the knowledgeable advisor as part of the iterative MLS cycles. Knowledgeable advisor involvement during the iterative cycles provided a form of formative assessment as the prospective teachers trialed and revised their emerging knowledge and skills in practice.

The prospective teachers engaged in the MLS also credited the videotapes of their lessons as providing an important form of feedback. Comments such as the following about analyzing one another's lesson videos were found repeatedly in the data: "This [analyzing one another's videos] allowed for feedback from others to assist in our growth and improvement of the lesson." Comments about the videotaped lessons also revealed the value felt by the prospective teachers for the opportunity to reflect on their own teaching. One such comment was, "The video itself. 'Seeing' your teaching is 'perfect memory' and allows for better reflection." This finding is in accord with Kpanja's (2001) suggestion that the use of videotapes has potential for providing prospective teachers with more complete feedback on their teaching. The videotapes allowed the prospective teachers to review their teaching more closely and even when their MLS group members' analyses lacked depth, the prospective teachers could reflect more deeply on their own teaching through use of the videotapes. As one prospective teacher commented after watching her videotape, "I believe that I could have been a little more prepared with the definitions and paid a little more attention to the students constructing their ideas of the concepts but I don't think that John and Tami [her MLS group members] would say that to me." The videotaped lessons were also beneficial for the instructor to watch when meeting with an MLS group to discuss one of their lessons, since the instructor often observed only part of each concurrent MLS lesson during. The videotaped lessons provided opportunities for formative assessment by peers, oneself, and the instructor.

For the prospective elementary school teachers, the feedback during the oral reports from knowledgeable advisors (course instructors) and other members of their methods classes provided for a summative form of assessment rather than formative assessment. For the prospective elementary school teachers, feedback from their lesson study group members during the iterative cycles was the only form of formative assessment that all groups were privy to. For the prospective elementary school teachers, negotiations with their group members sometimes led to a focus on issues of classroom management resulting in iterations of a lesson that were weaker in the depth of the mathematical ideas for student learning. For example, one group eliminated the use of hands-on materials for exploring a mathematical idea because the use of the blocks was new to the children and according to the prospective teachers caused disruptive behaviors. In another instance, due to similar reasons, a lesson study group revised a lesson by eliminating the group work. In situations such as these, participation by a knowledgeable advisor (e.g., course instructor) in the analyses of group lessons could have helped focus the prospective teachers' attention on the students' learning of the mathematics (Fernandez 2009) and mitigate the teachers' focus on issues of classroom management which have been found to be easy lures for novice teachers' attention when thinking about teaching (Moore 2003; Zeichner and Tabachnick 1981). The MLS groups discussed some general pedagogical issues related to classroom processes and management (e.g., ways of keeping students on task, lesson transitions, providing clear directions, time management, etc.) at times in order to address problems with their student-peers' behavior including writing personal notes, reviewing their personal calendars, chatting with peers and demonstrating other off-task behaviors. Nevertheless, these pedagogical issues were tied to discussions of their student-peers' learning of the mathematics, a focus encouraged and supported by interactions with the course instructor.

Community-Centeredness of the Lesson Study Approaches

The structure-of-the-experience design elements of the lesson study approaches contributed to the community-centeredness of each approach, supporting prospective teachers' learning from one another, striving for improvement, and feeling connected to one another and the larger community. For both approaches, during their group planning, analyses, and revisions, the prospective teachers often contributed differing instructional approaches, understandings of the mathematics, and perspectives for teaching the mathematics. The sharing and negotiation supported their learning from and building onto one another's ideas. The prospective secondary school teachers were organized into MLS groups heterogeneously with respect to their perspectives on teaching mathematics and their mathematics knowledge and understanding. Many of them commented on the importance of learning from one another through the MLS experience, making comments such as, "Working with my group members helped me consider ideas for teaching the lesson that I did not think of." Analysis of their group members' videotaped lessons was also important toward the community-centeredness of the MLS approach by fostering feelings of connectedness with one another and possibly the broader mathematics teaching community. Such comments included, "By analyzing one another's videos, I got to see that issues I had aren't necessarily isolated to me."

The prospective elementary school teachers felt that working in their lesson study groups was beneficial for their learning to work with other teachers. Many expressed sentiments such as, "What I got most out of this was learning to work with others. I learned to work with different personalities along with other teaching beliefs and styles." Their remarks also suggested the merits of listening and learning from other teachers, as suggested by the following:

> I felt this Lesson Study was difficult at times but made me realize that this is just the beginning because in the future I will be collaborating with others who may not have the same beliefs as me. Therefore, this whole Lesson Study was an eye opener to me and has taught me to keep an open mind and listen to what others have to contribute to the team.

The oral reports that were part of the elementary prospective teachers' lesson study experience also contributed to the community-centeredness aspect of their learning. The oral reports provided opportunities to help the novices understand how their peers in the mathematics methods classes thought about their lessons and the kinds of changes that they made in their lesson plans across the iterative cycles. The prospective teachers were encouraged to raise issues and questions during the oral reports and share alternative perspectives and revisions related to the lessons discussed. These discussions were beneficial in helping lesson study groups learn from one another by rethinking some of their naïve revisions such as eliminating the use of group work or hands-on manipulatives because of student disruptions. Insights from other groups that initially had similar student disruptions but made modifications in ways that successfully maintained the use of cooperative groups or manipulatives for promoting mathematical understanding helped the lesson study groups consider alternatives to their own solutions. The oral presentations also helped the elementary prospective teachers realize that many of their struggles as beginning teachers were experiences shared by others, thus precipitating feelings of connectedness to one another and the larger community of beginning teachers. In some cases, when the instructor knew how a group had successfully addressed a common problem under discussion, he would ask that group to explain how they handled the situation. In this way, the instructor provided direction for the discussion in order to offer opportunities for the prospective teachers to continue thinking about and learning from one another about teaching mathematics.

Discussion

Findings from this investigation support the value of these lesson study approaches in providing opportunities for prospective teachers to enhance their understanding of teaching mathematics. The structure-of-the-experience design elements that promoted learner-centeredness and the collaboration within the cooperative lesson study groups that promoted community-centeredness provided opportunities for learning from teaching that are not traditionally availed to prospective teachers of mathematics within traditional teacher education coursework and field experiences. Both within the secondary student-peer and the elementary school-based implementation settings, the prospective teachers were able to trial and develop, in a disciplined way, their emerging understanding of practices for teaching mathematics that challenge students' thinking and engage them actively in learning (NCTM 2000; AAMT 2002). The incorporation of lesson study approaches in our work with prospective teachers helped to lessen the idiosyncratic and particular character of learning to teach mathematics that was described by Ball and Cohen (1999). It brought prospective teachers together in cooperative groups to collaborate on the development, teaching, analysis, and ongoing revision of mathematics lessons, exploring in practice the implementation of their prior knowledge and their emerging understanding of the mathematics education reform-oriented ideas under discussion in their methods courses. The cooperation and negotiation related to the shared teaching experiences positively influenced the prospective teachers' learning about teaching and student learning, backing the claims that collaboration and shared experiences are important conditions supporting prospective teacher-learning (Putnam and Borko 2000).

At times, within the present investigation, collaboration with peers did not seem to bridge a group's knowledge and understanding of teaching mathematics toward reform-oriented practices. This was particularly the case in a few elementary teacher groups who did not have an opportunity to interact with a knowledgeable advisor (e.g., course instructor) during their actual teaching cycles. The differences in the forms-of-feedback design elements between the secondary and elementary school teacher groups seemed to play an important role in this finding. The forms-of-feedback elements contributed to the assessment-centeredness of the lesson study approaches. With respect to the MLS, the course instructor's discussions with lesson study group members during the MLS cycles and the use of videotaped lessons for self-reflection as well as peer and instructor analysis during the cycles contributed to the effectiveness of the MLS approach for prospective teacher learning. Both forms of feedback were availed to the prospective teachers in the midst of the lesson study cycles when they had opportunities to revise and retrial their research lessons. Such formative feedback during the lesson study cycles by individuals with pertinent mathematics teaching expertise (e.g., course instructor) can help bridge the prospective teachers' ZPD (Vygotsky 1978) and is said to add to the opportunity for learning by prospective teachers within authentic instructional tasks (Iverson et al. 2008; Darling-Hammond and Snyder 2000). Knowledgeable advisors that are part of the formative assessment aspect of lesson study experiences for prospective teachers can help focus the prospective teachers' attention on the student learning of the mathematics in ways that supports a group's trial and adjustment of practices aligned with recent mathematics education reforms and away from a focus on classroom management without consideration of the mathematics being learned by the students.

The videotaped lessons also contributed to the formative assessment aspect of the lesson study, allowing for prospective teachers' self-assessment of their teaching of their group research lesson and comparison of their teaching to that of their MLS colleagues. The videos also allowed for the prospective teachers to compare their teaching to that of other class members who taught them mathematics as part of the MLS. Korthagen et al. (2006) found that reflection on their own teaching and that of their peers in a peer teaching setting was valuable for prospective teacher learning. Self-reflection is important for teachers and is therefore essential for authenticity in approaches to teacher education (Darling-Hammond and Snyder 2000; Iverson et al. 2008). The videotapes were used "iteratively to inform practice," an important attribute of self-reflection in teacher education instructional tasks (Iverson et al. 2008).

The oral reports given by the elementary school lesson study groups to their methods classes including their instructors provided opportunities for the groups to receive feedback from their instructors, as well as from peers in other groups that might have handled similar difficulties in more reform-oriented ways. This reporting provided opportunities for the prospective teachers to continue thinking about their mathematics teaching; however, this type of feedback was not iteratively linked to the cycles informing their learning from practice. Thus, it may not have the same impact on prospective teachers' learning about teaching mathematics as may be the case with formative feedback from a knowledgeable advisor that is linked to cycles of revision and retrial in practice, as with MLS (Fernandez 2009). Formative feedback from knowledgeable advisors during the iterative lesson study cycles contributed to the assessment-centeredness of the experience and, in turn, to the increased effectiveness of lesson study approaches for prospective mathematics teacher learning. Reporting orally on lesson study experiences seemed to contribute more substantively to the community-centeredness of the experiences for promoting prospective mathematics teacher learning.

The lesson-content design elements of the lesson study approaches contributed to the knowledge-centeredness of the prospective teachers' learning experiences. These elements, particularly the overarching learning goals for the lessons, helped the prospective teachers focus attention on what they were learning in their coursework about the discipline of mathematics education. The prospective teachers expressed appreciation for the connection between theory and practice facilitated by the lesson study approaches, as indicated in the following representative comment: "We spend a lot of time discussing theory. It's good to get a chance to try using some of it and get feedback on how well we did before we get our own classrooms. More activities like this would be helpful." Fernandez and Robinson (2007) found that prospective teachers overwhelmingly perceived a lesson study approach (specifically, MLS) as important to their learning by helping them connect theory to practice. The lesson study approaches we investigated helped us, as teacher educators, better meet the calls for developing pedagogical approaches that challenge prospective teachers' thinking about teaching mathematics and aide in their understanding of reform-oriented teaching and connections of theory to practice (Graeber 1999; Grossman 2005; Bransford et al. 2000). In addition, the MLS supported the prospective secondary school teachers' development of knowledge of mathematics content and mathematical processes through their sense-making of mathematics for which they lacked familiarity or understanding.

Through the lesson study approaches, the prospective teachers developed their knowledge of mathematics pedagogy, mathematics content, and general pedagogy to different extents dictated, at least in part, by the interactions, negotiations, and deliberation-in-process (Fernandez 2009) within their groups. Both lesson study approaches required the prospective teachers to collaborate with a small cooperative group of peers during the lesson study cycles culminating in their creation of group reflective reports (written and oral) documenting their lesson study work. The creation of these reports formed the basis of a group goal necessitating negotiation and collaboration within the lesson study groups and guiding individual and group accountability, all key elements of effective cooperative learning (Johnson et al. 1994). Devoid of such a group goal and other related cooperative learning elements, the prospective teachers may have only partly collaborated, taking ideas as desired from one another's suggestions when preparing for their own teaching of the lesson, without engaging in a disciplined exploration of the evolution of their research lessons which is important to the learner-centeredness of the experiences. From this perspective, the production of written reflective reports on their lesson study work demonstrating growth in understanding teaching for the overarching mathematical student learning goals, which is an important feature of Japanese lesson study (Fernandez and Chokshi 2002), appeared to be essential with respect to the learner-centeredness of the lesson study approaches.

Comparisons across the two lesson study approaches revealed positive effects from engaging prospective teachers in authentic learning experiences that are more broadly defined than individual school-based classroom situations in real time. Both experiences engaged the prospective teachers in the kinds of thinking and problem solving important to teaching (Putnam and Borko 2000) through communication and reflection on shared contexts, one a typical classroom setting and the other a student-peer small class setting. Although the secondary prospective teachers were teaching student-peers, they were authentically teaching the mathematics to their small classes and were focused on student learning while at the same time dealing with issues of classroom management. Based on the prospective teachers' learning, what was simplified did not appear to be what makes teaching mathematics difficult, a goal proposed by Grossman and McDonald (2008) in rethinking approaches such as microteaching toward the development of "intensive, focused opportunities [for prospective teachers] to experiment with aspects of practice and then learn from that experience" (p. 189–190). Even though the MLS only partly addressed the complexities of teaching mathematics in a secondary school classroom, the experience seemed to help challenge the secondary prospective teachers' 'apprenticeship of observation' (Lortie 1975). The experience fostered the development of their understanding of recent reforms in mathematics education and provoked changes in their teaching and views toward more student-centered practices.

Concluding Remarks

As mathematics teacher educators, our intent was not for prospective teachers to develop a set of 'best' mathematics lessons, but to invite them to participate in the professional development process of developing, analyzing and discussing lessons. As researchers, we found the lesson study approaches helped our elementary and secondary prospective mathematics teachers to better understand and begin implementing teaching practices that are consistent with reform-oriented mathematics teaching. The experiences provided them with opportunities to expose their beliefs and practices to the scrutiny of peers and other knowledgeable advisors, trialing and revising their beliefs and practices within the iterative lesson study cycles. The lesson study approaches in teacher education provided our prospective teachers with opportunities to develop and teach lessons differently than they themselves might have been taught, or they might see taught in field experiences. The lesson study approaches helped provide them with a sense of ownership and control over lessons and the mandate, or at least license, to make and trial changes. The prospective teachers began to consider lessons as "works in progress" and they looked forward to "working with other teachers in the future." The decision-making about teaching mathematics was placed more in their hands, as a negotiated process rather than an isolated endeavor.

In educational contexts where teachers are being disenfranchised from the enterprise of making curricular decisions and developing appropriate instructional activities for their students, sometimes through directives to use prescribed script-based curricula, the implementation of lesson study experiences in teacher education may take on additional importance. Our lesson study approaches arguably could provide our prospective teachers with a view of teachers as researchers and decision-makers in their own classrooms rather than teachers as technicians who implement prescribed scripts and texts, a view promoted by elements of the current educational climate in some contexts. The lesson study approaches portrayed for our prospective teachers the view that quality mathematics instruction can and does come from classroom teachers, including themselves.

References

Australian Association of Mathematics Teachers (AAMT). (2002). *Standards for excellence in teaching mathematics in Australian schools*. http://www.aamt.edu.au. Accessed January 2005.

Ball, D. L., & Cohen, D. (1999). Developing practice, developing practitioners: Toward a practice-based theory of professional education. In L. Darling-Hammond & G. Sykes (Eds.), *Teaching as the learning profession: Handbook of policy and practice* (pp. 3–32). San Francisco: Jossey-Bass.

Bransford, J., Brown, A. L., Cocking, R. R., & CBSSE. (2000). *How people learn: Brain, mind experience, and school*. Washington: National Academy Press.

Cavin, R. (2007). Developing technological pedagogical content knowledge in preservice teachers through microteaching lesson study. Unpublished doctoral dissertation. Florida State University, Tallahassee.

Darling-Hammond, L., & Snyder, J. (2000). Authentic assessment of teaching in context. *Teaching and Teacher Education, 16*(5–6), 523–545.

Denzin, N. (1984). *The research act*. Englewood Cliffs: Prentice Hall.

Fernandez, C., & Chokshi, S. (2002). A practical guide to translating lesson study for a U.S. setting. *Phi Delta Kappan, 84*(2), 128–135.

Fernandez, M. L. (2005). Learning through microteaching lesson study in teacher education. *Action in Teacher Education, 26*(4), 37–47.

Fernandez, M. L. (2010). Investigating how and what prospective teachers learn through microteaching lesson study. *Teaching and Teacher Education, 26*(2), 351–562.

Fernandez, M. L., & Robinson, M. (2007). Prospective teachers' perspectives on microteaching lesson study. *Education, 127*(2), 203–215.

Frykholm, J. A. (1998). Beyond supervision: Learning to teach mathematics in community. *Teaching and Teacher Education, 14*(3), 305–322.

Garet, M. S., Porter, A. C., Desimone, L., Birman, B. F., & Yoon, K. S. (2001). What makes professional development effective? Results from a national sample of teachers. *American Educational Research Journal, 38*(4), 915–945.

Graeber, A. O. (1999). Forms of knowing mathematics: What preservice teachers should learn. *Educational Studies in Mathematics, 38*, 189–208.

Grossman, P. (2005). Research on pedagogical approaches in teacher education. In M. Cochran-Smith & K. M. Zeichner (Eds.), *Studying Teacher Education* (pp. 425–476). Washington: American Educational Research Association.

Grossman, P., & McDonald, M. (2008). Back to the future: Directions for research in teaching and teacher education. *American Educational Research Journal, 45*, 184–205.

Iverson, H. L., Lewis, M. A., & Talbot III, R. M. (2008). Building a framework for determining the authenticity of instructional tasks within teacher education programs. *Teaching and Teacher Education, 24*, 290–302.

Jacobs, J. K., Hiebert, J., Givvin, K. B., Hollingsworth, H., Granier, H., & Wearne, D. (2006). Does eighth grade mathematics teaching in the United States align with the NCTM Standards? Results from the TIMSS 1995 and 1999 video studies. *Journal for Research in Mathematics Education, 37*, 5–32.

Johnson, D. W., Johnson, R. T., & Holubec, E. J. (1994). *Cooperative learning in the classroom*. Alexandria: ACSD.

Korthagen, F., Loughran, J., & Russell, T. (2006). Developing fundamental principles for teacher education programs and practices. *Teaching and Teacher Education, 22*, 1020–1041.

Kpanja, E. (2001). A study of the effects of video tape recording in microteaching training. *British Journal of Educational Technology, 32*(4), 483–486.

Lewis, C., Perry, R., & Murata, A. (2006). How should research contribute to instructional improvement? The case of lesson study. *Educational Researcher, 35*(3), 3–14.

Lortie, D. (1975). *Schoolteacher: A sociological study*. Chicago: University of Chicago Press.

Marble, S. (2006). Learning to teach through lesson study. *Action in Teacher Education, 28*(3), 86–96.

Moore, R. (2003). Re-examining the field experiences of preservice teachers. *Journal of Teacher Education, 54*(1), 31–42.

National Council of Teachers of Mathematics (NCTM). (2000). *Principles and standards for school mathematics*. Reston: NCTM.

Putnam, R. T., & Borko, H. (2000). What do new views of knowledge and thinking have to say about research on teacher learning. *Educational Researcher, 29*(1), 4–15.

Stigler, J. W., & Hiebert, J. (1999). *The teaching gap: Best ideas from the world's teachers for improving education in classrooms*. New York: Free Press.

Vygotsky, L. (1978). *Mind in society* (trans: M. Cole, S. Scribner, V. John-Steiner, & E. Souberman). Cambridge: Harvard University (Original work published 1934).

Zeichner, K. M., & Tabachnick, B. R. (1981). Are effects of university teacher education washed out by school experiences? *Journal of Teacher Education, 32*, 7–11.

Lesson Study in Preservice Elementary Mathematics Methods Courses: Connecting Emerging Practice and Understanding

Aki Murata and Bindu E. Pothen

Introduction

Teacher education programs have long struggled to find the best way to prepare quality teachers, and research has shown that teacher educators face several challenges. Lortie (1975) uses the term "apprenticeship of observation" in explaining how teachers' own schooling experiences shape their beliefs about teaching, and ultimately how they interact with students in their classrooms. Brouwer and Korthagen (2005) stipulate that teachers become socialized into the profession, adopting more traditional models of teaching that are consistent with the culture at most schools, despite the kind of experiences they have had in teacher education programs. We must question how much influence the teacher education programs can have on teachers given that such socialization is a strong factor impacting practice.

In addition, there is a body of literature that suggests that the lack of subject matter preparation also influences teaching (Ball 1990; Borko 1992). Researchers recommend that teacher education program leaders reconsider how subject matter knowledge is included in courses in a manner that challenges preservice teachers' beliefs. For example, preservice teachers may engage in collaborative activities that require them to reflect on how different concepts work together in problem solving, thus they will rethink and strengthen conceptual understanding while experiencing a collaborative learning environment.

There is evidence that teacher education can and does make an impact on some teachers' practice. Darling-Hammond (2006) found links between what was taught in teacher education programs and the strategies teachers employed in their classrooms. In fact, several studies have highlighted how methods courses that focus teachers' attention on building knowledge and a variety of pedagogical strategies result in greater ability to implement reform efforts espoused by teacher educators (Cunliffe 1994; McDevitt et al. 1999). Specifically in mathematics and science, there is evidence of a connection between teachers' acquisition of new teaching

A. Murata (✉)
Stanford University, Stanford, CA, USA
e-mail: akimura@stanford.edu

L. C. Hart et al. (eds.), *Lesson Study Research and Practice in Mathematics Education*,
DOI 10.1007/978-90-481-9941-9_8, © Springer Science+Business Media B.V. 2011

practices and their mathematics methods courses pedagogy (Judson and Sawada 2001).

It is also plausible that the learning from teacher education programs sometimes takes years to appear in practice. This may be linked to preservice teachers' lack of exposure to students, making it a challenge for them to understand how content and pedagogy connect to student learning. Placement in schools, with the aid of an experienced cooperating teacher, can work to address this problem in part. Still, scaffolded examinations of student thinking can also be helpful as preservice teachers make sense of their emerging practice.

Lesson study can be a cohesive professional development tool when faced with the challenge of providing high-quality learning experiences for preservice teachers. Preservice teachers are typically exposed to many new experiences in their teacher education programs. The two main contexts for these new experiences, teacher education course work and field placements, are often considered to be at odds (thus creating a gap between beliefs and practice). When there does not appear to be direct and obvious connections between what is learned in the course work and what is experienced in field placements, preservice teachers may prioritize what they learn in the field, as this learning is directly transferable to what they need to do in the new teaching context. As discussed in other chapters of this book, the process of lesson study that meaningfully connects different teaching activities (e.g., planning, assessing) can help bridge the gap for preservice teachers while keeping their focus on student learning.

Since the start of our elementary teacher education program at Stanford University in 2005[1], we incorporated lesson study to support our preservice teachers' learning of mathematics teaching. The elementary mathematics methods and lesson study process have been integrated in the courses and continually been modified and refined over the past years. Fernandez (2005) argued that lesson study could serve to focus teachers' attention on the critical work of student learning. Our preservice teachers are carefully guided through the lesson study cycle to deepen their understanding of student learning of mathematics. In this paper, we will outline the structures of our preservice mathematics methods courses with lesson study.

Lesson Study Course Activities by Week

In the initial planning of our preservice elementary mathematics methods courses, we tried to be true to the original Japanese lesson study structure (see the chapter by Meyer and Wilkerson, this volume) as closely as possible. We also made appropriate adjustments to meet the particular needs of the preservice teachers and

[1] *STEP* (*Stanford Teacher Education Program*)—*Elementary* is a twelve-month program that leads to a teaching credential and a Master of Arts degree. For more information: http://www.stanford.edu/group/step/new-step-website/elementary/index.htm.

the teacher education program. We have three quarter-long mathematics methods courses in the yearlong program, which are taught over three academic quarters (ten weeks each). For the entire program, the preservice teachers are placed in elementary school classrooms, while their responsibilities in the classrooms change over time. For example, their primary responsibility for the first few months will be observing and learning from being a part of the classrooms, and that will change to being responsible for teaching mini lessons with small groups of students by the middle of the year, to planning and assessing entire instructional units at the end of the year. For the first and second courses of the sequence, through activities, readings, and discussions, our preservice teachers begin to see teaching as interactive processes between teachers and students, recognize the importance of ongoing informal assessment of student thinking, and understand how to teach lessons with open-ended problem solving. In addition to this important pedagogical learning, teachers also begin to engage in investigations of elementary mathematics content. The primary activity in the third course is lesson study (the focus of this chapter), and the preservice teachers work collaboratively on lesson study for ten weeks. Table 1 is the lesson study calendar for the ten-week quarter.

One of the goals of this course is to facilitate integration of pedagogical strategies with new and deepened knowledge for teaching (resulting in a greater knowledge base), and subsequently using the knowledge to teach mathematics in classrooms.

In the following section, we describe the course, week by week, focusing on course activities and assignments. While students engage in lesson study activities, we also have time in class sessions for mathematics content activities and discussions. We usually allocate half of the three-hour class meeting time each week (approximately 1 h and 30 min) for lesson-study-related activities. Most assignments are group assignments, following the lesson study collaboration format. Our preservice teachers gain valuable experience working in groups, and this group format also reduces the number of writing assignments required by each teacher.

Table 1 Lesson study calendar

Wk	Lesson study topics and activities	Lesson study assignments
1	Getting overview of lesson study process	
2	Discussing cooperating-teacher interview results Deciding on lesson study goals and topics	Cooperating-teacher interview results
3	Situating goals with standards	
4	Situating student learning in research and deciding on assessment tasks	
5	Analyzing preliminary student learning data	Student assessment data
6	Planning the unit and lessons	Student assessment summary
7		
8		
	Research lessons (taught in Weeks 8–9)	
9	Reflecting as a group	Unit/lesson plans
10	Sharing lesson study research experiences in whole-class presentations	Presentation and lesson study portfolio Reflection paper

Week 1: Introduction to Lesson Study

Our preservice teachers come to the first class session having completed assigned readings on lesson study (overview of lesson study with some practical example of how it is used in schools: e.g., Lewis and Tsuchida 1998). As we give the overview of the course, we focus our discussion on the lesson study process and emphasize how it will work to help connect the different parts of teaching that they have experienced so far into a meaningful whole. The preservice teachers are teamed according to their grade level field placements, each group consisting of 3–5 teachers. After watching a video of a lesson study case (*How many seats?* Mills College Lesson Study Group 2004), we end the class by giving guidelines for the upcoming assignment. In this first assignment, the preservice teachers interview their cooperating teachers about "challenging" mathematics topics to teach, specific to their grade levels. This is a crucial assignment as preservice teachers often lack experiences to determine gaps in students' mathematical knowledge, and how to decide on the relevant mathematics topic. Thus, their cooperating teachers' knowledge plays an important role in their development. This assignment also helps maintain the connection between field placement classrooms and university courses.

Week 2: Deciding Lesson Study Goals and Topics

The preservice teachers share their interview results in their lesson study groups. We often limit the content strand for the interview (e.g., geometry, number sense) so that we can anticipate likely topics cooperating teachers might choose. We ask cooperating teachers to identify the top three challenging topics in the content strand, thus increasing the possibility for common topics across teachers. The preservice teachers share their interview results, negotiate, and decide on the topics for their lesson study effort for the course. They also discuss the challenges that their cooperating teachers identified in student learning of the topics. In their groups, they discuss and make a list of what they currently know about the topic and student learning of the topic. This list is kept so that they will be able to reflect on their learning at the end of the quarter. Preservice teachers may also identify "social" goals for their students that can range widely: e.g., "working collaboratively in groups," "having each student speak at least once in class discussions." For each lesson study meeting, each group identifies a facilitator who focuses the group's discussion on given topics and a note taker who keeps a written record of the discussion.

Week 3: Situating Goals with Standards

While this is not a priority in working with inservice teachers (primarily because they do this regularly), it is critical for preservice teachers to understand and become familiar with content standards. For the Week-3 meeting, teachers examine both state and NCTM (National Council of Teachers of Mathematics) standards, locate where their chosen topic fits in the standards, and identify how the standard is related to the standards in the previous and subsequent grade levels. Each lesson study group creates a graphic organizer to illustrate the relationships among standards and grade level expectations.

Week 4: Situating Student Learning in Research and Deciding on Assessment

Prior to this class meeting, each lesson study group is provided with appropriate research-related readings on their chosen mathematical topic (instructors prepare the readings). In the meeting, the teachers discuss research on typical learning trajectories of the topic and possible challenges students experience in the process. They then discuss and decide, based on the readings, assessment items to administer to their students in order to understand their current level of understanding of the topic. Please see Appendix 1 for the assessment assignment guidelines[2].

Week 5: Analyzing Student Data

The preservice teachers bring student data from their preassessment interviews. After sharing what they saw/found in the data, they discuss if there are patterns in terms of student strategies and mistakes. They return to the research literature from the previous week to make sense of the patterns. Working in teams, they are likely to see similarities among students at a grade level. They create tables and charts to organize the data as a team, write a short paper discussing the patterns they see, and tie the results to the literature.

Weeks 6–8: Unit and Lesson Planning

We allocate three weeks for lesson planning. The teachers are advised that they should find time outside of class to meet and work on their lesson plans at this point.

[2] In the first and second math courses of the three-quarter sequence, the teachers have conducted informal assessment involving student interviews. Therefore, they are familiar with the assessment procedures and expectations. Preservice teachers often find it challenging to take notes from what students say and how to identify student thinking in their explanations. Reading research materials helps, but the teachers need to spend some time practicing and brainstorming possible interview settings prior to the actual face-to-face student interview experiences (e.g., watching student interview videos and discussing student strategies, role playing with peers).

We typically provide several published lesson plans or reform-based curricular materials (e.g., Gavin et al. 2001) for the teachers and strongly encourage them to start from these existing materials as they plan lessons, and not try to create lessons from scratch. We explain that as beginning teachers, they are likely to work with a specific curriculum/textbook provided by the district/school once in classrooms, and modifying and adjusting existing materials for their own students is a skill they need to learn. By this point, each lesson study team has decided which preservice teacher is teaching the research lesson. That teacher provides necessary information about his/her students to the team (to aid in adjusting the lesson).

The teachers are provided with a lesson plan format, which is specifically designed for research lesson events (see Appendix 2).

The lesson plan has sections where the teachers insert information from course assignments thus far: goals of the lesson, relationships with standards, about the students (student preassessment), about the lesson (how they planned the lesson), and lesson steps. The lesson steps table (Appendix 2, #9) is made up of three columns: one for teacher activities, one for student activities, and the last one for evaluation points. While most other lesson plan formats focus only on what teachers do in the lesson, this format encourages the teachers to think about what students would do in relationship to what the teacher does. It also focuses their attention on particular evaluation points that will later help focus observers' attention on what to look for as they collect data on student learning during research lessons. The entire lesson plan format pulls together what they have done so far in the course (from various assignments and activities) to present their efforts into a coherent picture.

We ask the teachers to think deeply about what they have learned so far and to generate "research questions" for their lesson study. The questions should be closely tied to student learning of the topic they chose and the learning goals they specified at the beginning, but need to be refocused and more specific. For example, a lesson study group may be planning a lesson for simple multiplication word problems for Grade-2 students. NCTM Standards, Pre-K–2 Number and Operations, says "understand situations that entail multiplication and division, such as equal groupings of objects and sharing equally," and in the research lesson, the teachers may be interested in knowing what multiplication strategies Grade-2 students use to solve a division problem. Accordingly, they would specify a point in the lesson plan (on the third column) to collect data on student strategies when the key problem is presented to the students. In this way, the teachers continuously narrow their thinking on student learning of the topic and deeply consider certain student learning in the lesson, which will make data collection more meaningful.

Research Lesson Events

In consultation with the cooperating teacher whose classroom they use, we allow each lesson study team to decide when to teach their research lesson. The research lesson event can occur any time in Week 8 or Week 9. Prior to the lesson, the preservice teacher who is teaching the lesson makes arrangements with his/her cooperating teacher to set up the room for the lesson, inform the office staff and principal about the event, and communicate with students about what will be happening. Guidelines for the research lesson event are provided for the teachers (see Appendix 3). Course instructors personally invite principals, administrators, and district professional development staff to research lessons.

On the research lesson day, events unfold as summarized in the guidelines in Appendix 3. While most teachers may admit how nervous they feel anticipating the lesson, most of them report that they leave the experience feeling rewarded and connected with one another in the team. Course instructors facilitate the event. Primarily for those observers who are new to lesson study, we emphasize that our focus is to learn about student learning of mathematics. We ask observers to collect data that are specified in the lesson plan materials, and then discuss the data as evidence during the debriefing that follows the lesson. During the debriefing, when the discussion shifts away from student learning and other issues are discussed, we gently redirect the conversation so that we can focus once again on the student thinking observed during the lesson. In this way, we prevent the discussion from becoming a critique of pedagogy and the teacher. However, when teaching is discussed in relation to student learning during the lesson, we allow room for this type of discussion.

Week 9: Group Reflection

In the Week-9 class meeting, lesson study teams come together and reflect on the learning process they shared with one another and begin work on their final projects (see Appendix 4). The preservice teachers bring all their work samples from the course to this class, so that they have ready access to all of the materials they need to reflect. We allow the entire class period for this activity (three hours) to allow teachers enough time for group as well as personal reflection. We provide sample portfolios from previous years, large binders, and supplies necessary to put the portfolios together.

Week 10: End-of-Quarter Presentation

As described in the finals guidelines (Appendix 4), each lesson study group presents an account of the learning process to the whole class. We celebrate the collective hard work of the course.

Discussion

We use lesson study in our courses to support the connection-building and sense-making process for our preservice teachers, to support integration of knowledge and practice, and to inform their future teaching. Lesson study is one way to value the knowledge and data we bring from the field into our course work. This is achieved by maintaining strong connections between course activities and field-based assignments (e.g., cooperating-teacher interview, preassessment of student learning). We stress the importance of practice in our research and theory building, as some of our preservice teachers choose to become educational researchers and teacher educators in the future. This also serves to reinforce the professionalism of the teachers, reassuring them that they are important actors and participants in the research process.

In addition to facilitating the connection-building process between their course work and field experiences, lesson study can also serve as a mechanism through which preservice teachers connect various aspects of teaching that may seem discrete from one another. Through assignments in their teacher education courses, preservice teachers learn a variety of critical pedagogical skills. For example, in one assignment they may conduct student interviews to learn more about the value of assessment data. In another assignment, they read about problem-based teaching, watch a video of an exemplary problem-based classroom scenario, and discuss this structure. They are likely to consider these two course experiences separately, unless their attention is meaningfully focused on the fact that assessment of student learning drives problem-based teaching, and that teachers need to understand student thinking to guide productive discussion. The lesson study cycle connects these different teaching-related activities meaningfully as one activity builds on another.

As we delve into a next phase of the lesson study process, we continue to make references to the work that the preservice teachers have done so far in the course. We do this to make explicit connections among the different experiences. In the research lesson, it becomes obvious that all of the activities and assignments they have completed in the course are important parts of the culminating lesson. Creating a final portfolio provides teachers with an avenue to reflect on their work.

What makes this connection-building process unique in lesson study is its focus on student learning. In each part of the lesson study process, the teachers focus on how students learn the particular lesson study topic of their choice. For example, when they examine standards, they consider how different standards relate and connect from one grade level to the next to support student learning of the topic over time. When they conduct preassessment, they take ideas from literature to understand the thinking of the students in their elementary classrooms. When they plan lessons, they discuss how standards, research literature, preassessment results, and curricular materials connect in terms of student learning on the topic.

Through these experiences, the preservice teachers come to understand the importance of student learning in the lesson study process. Beginning teachers may view teaching as separate from student learning when they do not immediately see the impact of their work on student performance. They may feel lost if they believe that the lessons they teach are not effective. By understanding the connections between different aspects of their work, they are better able to look for resources in different places and shift their approach to support student learning.

The connections built between teaching, student learning, and ideas from theory/research support the development of knowledge for teachers. In lesson study, preservice teachers deepen their understanding of mathematics content through investigating student learning of the content. As they think more about how elementary students learn the mathematics topic, they reflect on their own thinking of the topic and plan research lessons. They are required to anticipate student responses to each question they pose in the lessons. This exercise challenges the teachers to revisit the content topics more deeply. Preservice teachers learn through practice how to guide student discussions, and their existing and developing knowledge supports effective facilitation of the discussion. The knowledge is meaningfully used in their classroom practice and further developed.

One aspect of lesson study which preservice teachers highly value is collaboration. Lesson study provides reasons for the teachers to collaborate meaningfully in the course, and some assignments can only be completed when all members of the group participate (e.g., student assessment). While some groups may take a longer time to develop productive working relationships, they come to appreciate each other's presence in the process. When research lessons are taught, many teachers mention how they recognize the value of having members of the lesson study team observe student learning in their classrooms. Lesson study allows teachers to stop and observe different learning "in the moment" with multiple eyes. For both inservice and preservice teachers, this can be a valuable experience through which they deepen their understanding of teaching and learning.

Lesson study, through its strong focus on collaborative work, works to counter both the socialization effects of schools and the "apprenticeship of observation." Teachers who favor a more traditional pedagogical approach may come together in collaboration with teachers who prefer a more reform-oriented approach, and in working together these teachers discuss their views and unite in a better approach for the students. While we support our elementary school students to work collaboratively to learn mathematics in classrooms, it is disappointing that teachers continue to work separately and independently from one another. We hope that through the lesson study process our preservice teachers take the desire to work with and learn from others to their future work environments. Through asking questions of their colleagues, collaboratively investigating student data, and simply sharing experiences at the end of the day, teachers will begin to have a shared responsibility for all students.

This chapter presents a description of the use of lesson study in an elementary preservice mathematics methods course, and illustrates its potential in supporting teacher learning and instructional improvement. As with other professional development formats, it is important to modify the process to meet the needs of those in a particular context. We are presenting our case to be one example, while recognizing that there may be other courses which, while different from ours, support their preservice teachers' learning just as effectively. As lesson study is a new territory in the United States and other parts of the world, sharing each group's efforts with others is valuable and we will surely benefit from continual development and refinement of our models.

Appendix 1: Student Assessment Assignment Guidelines

1. In your lesson study group, discuss key ideas for teaching and learning (Weeks 3/4)
    a. Why is this topic challenging to teach and learn?
    b. What are the typical mistakes students make?
2. Create a set of assessment items (Week 4)
    a. Start with what you discussed in the previous lesson study meeting. What are the key ideas and what is challenging?
    b. Design assessment items that highlight the difficulty and challenges of the topic (pay attention to materials used, order of questions, etc.). What does this assessment help you learn about your students' thinking? Anticipate your students' responses.
    c. Decide whether the questions will be interview questions or paper-and-pencil questions.
    d. Individual assessments should not last longer than 15 min (not too many questions!).
3. Assess student thinking (Week 4/5)
    a. The assessment guidelines and expectations have been communicated with your CTs.
    b. Ask your CTs to help you identify a student sample in the classroom (ideally 2 students who achieve beyond grade level, 2 right at the grade level, and 2 struggling at the level; include ELL student).
    c. If you choose to use paper-and-pencil format, find a time to talk with each individual student afterwards and ask him/her to explain thinking process.
4. Bring back assessment data to class (Week 5)
    a. You will discuss what you found in your lesson study group.
    b. What are common mistakes? Any pattern?
    c. Why such mistakes? [Connecting back to research]
    d. What do the results suggest for your teaching (regarding topic, question types, methods you might use, particular student needs)?
5. Write up assessment summary
    a. Attach all the assessment data with the summary.

Appendix 2: Lesson Study Lesson Plan

In writing a lesson plan, please include the following:

1. **Title of the Lesson**
2. **Names of the teachers**

3. **Grade level of the lesson**
4. **Goals/objectives of the lesson** [content objectives as well as social/affective and ELL objectives if appropriate]
5. **Relationship of the lesson to standards** [mathematics content standards for California public schools K-12, and the National Council of Teachers of Mathematics Standards]
6. **About the Lesson** [discuss how this lesson came about]
7. **Placement of the lesson in the unit** [where does this lesson fit in the unit you planned? Discuss the connections among lessons]
8. **About the students** [briefly discuss your student assessment results]
9. **Lesson steps** [use the three-column format—please see back of the page]

Teacher support [This column explains lesson flow]	**Student activity** [This column shows anticipated student responses to the lesson activities]	**Points of evaluation** [This column focus on particular aspect of the lesson]

10. **Evaluation** [this needs to correspond to the objectives you stated at the beginning of the lesson plan]

Appendix 3: What to Expect on the Research Lesson Day…

You have worked very hard with lesson study, and the research lesson is the opportunity to celebrate your accomplishment. You chose a challenging math topic to teach, read research literature to learn why it is challenging, interviewed elementary students to find out their strategies, examined different curricular materials to see the textbook approaches to the topic, and planned your own lesson. The lesson will make your ideas and learning come to life in classroom with real "live" kids. You are the expert now, and share your pride with others who come to observe!!

Before the lesson:

- Let office people know you are expecting visitors that day.
- Make multiple copies of lesson plans.
- Get all the lesson materials ready.
- Set a video camera in the back of the room and one group member can monitor the filming (if there are students whose parents do not wish them to be videotaped, move the students to the side of the room, or set the camera so that these children won't be in the picture).
- Decide who is giving the intro of the lesson (2-minute overview, see below) to the observers.
- Decide who is taking notes during the debriefing.

1. **Introduction of the lesson (start approximately 15 minutes before the lesson)** Observers gather in front of the classroom. Bring copies of lesson plans and give them to the observers. One member from the planning team (usually not the instructor) gives a brief overview of what will happen during the lesson and talk about the data they would like observers to collect during the lesson (it should be noted in the lesson plan). The team can also speak to their reasoning behind decisions they made about the lesson structure. The facilitator gives brief guidelines on observation protocol (handout provided). Allow a few minutes for observers to look over the lesson plan.
2. **Research lesson (up to 60 minutes)** Enjoy the lesson. Support the instructor as appropriate. Resist the urge to help students or otherwise interfere with the lesson.
3. **Debriefing (approximately 45 minutes)** Find a space for the lesson debriefing (staff room, empty classroom, bench at playground, nearby coffee shop, etc.). Decide one group member to be the scribe and keep notes of the discussion. The debriefing will follow the steps:

   a. **Instructor's reflections:** The instructor describes his/her experiences in the lesson and comments on what went well and was unexpected.
   b. **Planning group members' comment on the data gathered:** This may include data on student learning and engagement. Comments may be guided by two or three specific questions related to the goals of the lesson.
   c. **General discussion on the data gathered:** A brief discussion period, facilitated by a moderator, focused on student learning and development, and how specific elements of the lesson promoted these.
   d. **Outside commentator:** An outside expert will comment on mathematics and student learning of mathematics.
   e. **Thanks:** The facilitator makes final comments and thanks everyone for the participation.

Appendix 4: Final Project Guidelines

Your finals consist with two parts: Lesson study group portfolio and group presentation.

1. Lesson study group portfolio: You will put together a portfolio as a lesson study group. Portfolio is not a mere collection of items but should be organized with coherent themes to guide your reflection. Your portfolio will have following categories:

   a. The work we learned the most from
   b. The work we had the most fun with
   c. The most challenging work
   d. The least favorite work
   e. (Your own category)

For each of the categories, your group will choose one item from lesson-study-related work you have done this quarter. Your group may also create your own item for a category. For example, you may say "students' math talk during the lesson" as the most favorite part of your lesson study experience, and in that case, you can put some representation of the experience (photo, etc.) in the binder for the category and write a short description.
In the big binder provided, label a divider with each category, include item(s) that you chose for the category, and write a short paragraph stating WHY you chose the particular item for the category. For the assignment you did not select, add them at the end of the binder. Include all the lesson-study-related materials in the binder (e.g., copies of meeting notes, debriefing notes, student work from the lesson). Please also turn in your lesson CD with the portfolio.

Individual Reflection Papers (to be included in the portfolio): While the portfolio is assembled, reflect on the lesson study process and your learning in the quarter. Write a short individual reflection paper and include at the very end of the portfolio. The reflection paper should answer the following questions (but not limited to them):

- What do you think you learned through the lesson study process?
- What ideas from the lesson study experience do you think you can take to your future teaching?

2. Final lesson study group presentation:
Prepare a group presentation to highlight your learning with the lesson study in the quarter. Each group's presentation should be approximately 20 minutes, and I encourage you to use some visuals—powerpoint, video, overhead, handouts, etc. Please include presentation materials in your final portfolio (e.g., copy of handout, copy of powerpoint slides). I can naturally see how you may use the portfolio categories to organize your presentations, but it is not required. Be creative and have fun with it☺!!

References

Ball, D. (1990). The mathematical understandings that prospective teachers bring to teacher education. *Elementary School Journal, 90*(4), 449–466.

Borko, H., Eisenhart, M., Brown, C. A., Underhill, R. G., Jones, D., & Agard, P. C. (1992). Learning to teach hard mathematics: Do novice teachers and their instructors give up too easily? *Journal for Research in Mathematics Education, 23*(3), 194–222.

Brouwer, N., & Korthagen, F. (2005). Can teacher education make a difference? *American Educational Research Journal, 42*, 153–224.

Cunliffe, A. (1994, July). *How do science teachers become professionals? Implications of case studies of two beginning teachers.* Paper presented at the annual meeting of the Australian Education Association, Brisbane, Queensland, Australia.

Darling-Hammond, L. (2006). *Powerful teacher education: Lessons from exemplary programs.* San Francisco: Josey Bass.

Fernandez, C. (2005). Lesson study: A means for elementary teachers to develop the knowledge of mathematics needed for reform-minded teaching? *Mathematical Thinking and Learning, 7*(4), 265–289.

Gavin, M. K., Pelkin, L. P., Spinello, A. M., & St. Marie, J. (2001). *Navigating through geometry in Grades 3–5.* Reston: National Council of Teachers of Mathematics.

Judson, E., & Sawada, D. (2001). *Tracking transfer of reform methodology from science and math college courses to the teaching style of beginning teachers of grades 5–12.* Paper presented at the annual meeting of the American Educational Research Association, Seattle.

Lewis, C., & Tsuchida, I. (1998). A lesson is like a swiftly flowing river: Research lessons and the improvement of Japanese education. *American Educator, 1998* (winter), 14–17 & 50–52.

Lortie, D. (1975). *Schoolteacher: A sociological study.* Chicago: University of Chicago Press.

McDevitt, T. M., Gardner, A. L., Shaklee, J. M., Bertholf, M. M., & Troyer, R. (1999). Science and mathematics instruction of beginning elementary teachers. *Journal of Science Teacher Education, 10*(3), 217–233.

Mills College Lesson Study Group. (2004). *How many seats?* (DVD). Oakland: Author.

Murata, A. (2010). Teacher learning with lesson study. In P. Peterson, E. Baker, & B. McGaw. (Eds.). *International Encyclopedia of Education, 7*, 575–581. Oxford: Elsevier.

Lesson Study as a Framework for Preservice Teachers' Early Field-Based Experiences

Paul W. D. Yu

Introduction

Initial field-based experience for preservice teachers often takes the form of observation of a classroom or tutoring a student. Later experiences may include assisting a teacher in the classroom, culminating in student teaching where the preservice teacher is in complete control of a group of students. Some have questioned the significance of field-based experiences as nothing more than an enculturation into the existing socio-cultural norms of the teaching profession (Zeichner 1981). As suggested in this volume (Murata and Pothen; Fernandez and Zilliox), one way to fold the myriad of preservice teaching activity into a meaningful field-based experience is to give preservice teachers an opportunity to conduct lesson study (Stafford-Plummer 2002).

The context for which lesson study was used in this study was a field-based experience early in a preservice teacher education program. The majority of the students were in their second or third year of college. The lesson study experience replaced what would typically be a series of classroom observations or tutoring experiences. This chapter describes the modification to lesson study used in this context in order to provide an alternative framework from typical early field-based experiences.

Philosophical and Practical Considerations

As is evident in this volume, the exact nature of lesson study is difficult to describe. There is no simple answer to the question "What is lesson study?" and how it is different from a "community of practice" (Grossman et al. 2001)? Furthermore, while identifying necessary components of lesson study is difficult, if not impossible,

P. W. D. Yu (✉)
Grand Valley State University, Allendale, MI, USA
e-mail: yupaul@gvsu.edu

L. C. Hart et al. (eds.), *Lesson Study Research and Practice in Mathematics Education*,
DOI 10.1007/978-90-481-9941-9_9,

some attempt to come to a shared description is important so as to have a way to evaluate modifications and uses in different contexts. So to help preservice teachers build a model of what lesson study is and how it is implemented, the article *Ideas for establishing lesson study communities* (Takahashi and Yoshida 2004) served as a primary resource. They describe the lesson study process as follows:

- Formulate long-term goals of student learning and development
- Plan, conduct, and observe a "research lesson" designed to bring these long-term goals to life, as well as to teach particular academic content
- Carefully observe student learning, engagement, and behavior during the lesson
- Discuss and revise the lesson and approach to instruction based on the observations (Lewis 2002 cited in Takahashi and Yoshida 2004, p. 436)

However, in spite of what may appear to be an agreed-upon-framework for lesson study, there are also important practical considerations between inservice and preservice teachers' professional contexts. For example, inservice teachers are in a classroom in which they meet with their school-aged students approximately six hours a day, 180 days a year. Preservice teachers attend college courses several hours a week for about 15 weeks. Also, inservice teachers have their own students. Preservice teachers have few, if any, connections to K-8 students. Finally, inservice teachers have some sense of what professional development means or looks like, while preservice teachers have little or no experience with professional development. Given these differences, it was necessary to adapt the lesson study model for preservice teachers.

This chapter describes how lesson study was used and modified in a preservice elementary mathematics education course and then explores the professional growth of the preservice elementary teachers' ability to plan a mathematics lesson. Specifically we will look at short case stories of a group of three students (Meghan, Mallory, and Mia) enrolled in this mathematics education course. These case stories are from transcripts of video footage taken of the three students as they participated in this lesson study project.

Overview of the College Course

The college course for which lesson study was utilized was the second of two required mathematics courses in which content and pedagogy were integrated, but before students take courses in the college of education. This 15-week course met two days a week for two hours each day. The two hours provided students with the time necessary to participate in investigative learning activities and accommodate some form of early field-based activity. At this point in their program, the college students' exposure to issues of assessment and pedagogy was minimal, and their mathematics knowledge necessary to teach was in the early stages of development. Two goals of the course were (1) to help students develop a deep conceptual understanding of mathematics, while exposing them to research-based pedagogy and (2) to provide them with a meaningful field-based experience to investigate children's understanding of mathematics while experimenting with the pedagogical principles that were presented in the course. Students in the course were put in working groups consisting of three or four students for the entire semester. The purpose was to create a classroom culture that was explicitly collaborative across all learning activities in this course. Thus, the expectation to investigate children's understanding of mathematics, experiment with new pedagogy, and work collaboratively in a lesson study were consistent with the students' overall experiences in the course.

The first eight weeks of the college course were spent developing the content and pedagogical content knowledge necessary to meaningfully participate in the lesson study project. For this particular project, the preservice teachers focused on whole-number operations using a Cognitively Guided Instruction (CGI) framework (Carpenter et al. 1999). Carpenter et. al. (1999) describe CGI as "a professional development program based on an integrated program of research on (a) the development of students' mathematical thinking; (b) instruction that influences that development; (c) teachers' knowledge and beliefs that influence their instructional practice; and (d) the way that teachers' knowledge, beliefs, and practices are influenced by their understanding of students' mathematical thinking." Since one goal of lesson study is to "carefully observe student learning, engagement, and behavior during the lesson" (Lewis 2002 cited in Takahashi and Yoshida 2004, p. 437), the use of CGI was a good fit to provide both a curricular framework for lesson planning and a theoretical lens for the analysis of the school-aged children's mathematical activity.

The Project Setting and Timeline

The lesson study project took place over a four-week period during the second half of a 15-week semester. The assessment of the children's understanding of mathematics, the teaching of the research lesson, and the research lesson debriefings took place in second- and third-grade classrooms at a local elementary school. All planning of the research lessons was done in the university classroom. Table 1 outlines the weekly timeline used for this project.

Each of the 24 students in the course was assigned an elementary school to pair up one-to-one with school-aged children. In week 1, the project began with each preservice teacher conducting a 60-minute mathematics assessment interview with their respective child. The purpose of the assessment was to help inform the preservice teachers as to the nature and extent of their assigned child's understanding of

Table 1 Lesson study schedule

	Tuesday—in the University classroom	Thursday—at a local elementary school
Week 1		Assess elementary children's mathematical understanding
Week 2	Research lesson 1 planning time	Research lesson 1 conducted and debriefed
Week 3	Research lesson 2 planning time	Research lesson 2 conducted and debriefed
Week 4	Research lesson 3 planning time	Research lesson 3 conducted and debriefed

mathematics in the area of whole-number operations and place value. The assessment was reflective of the CGI course content material, and was to be used in the formation of the learning goals for the lesson study. The use of the CGI framework for the course was the first modification to the lesson study process. While inservice teachers may have more latitude in determining the learning goals for their research lessons, the preservice teachers in this course were given the constraint of creating learning goals that utilize CGI based lessons, in the area of single-digit multiplication or division. The primary reason for this constraint was to create a unique teaching and learning experience for the preservice teachers. Since the elementary school classes were either second or third grade, the children in these classes usually had little to no formal experience with multiplication and division facts. Therefore, when given CGI multiplication and division word problems, almost all of the school-aged children had little to no formal fact or procedural knowledge to use in solving the problems. While not necessary, this constraint did provide a sense of uniformity of experience and focus for both the preservice teacher lesson study groups and the college instructor working to mediate the learning experience.

Lesson Planning

During week 2 of the project, in the next university class period following the assessment day, the preservice teachers were given 20–30 minutes to discuss in their lesson study groups what they had learned about their child's understanding of mathematics through the assessment interview. In the case of Megan, Mallory, and Mia (our example group), their discussion was centered around what each of their respective elementary students had done on the various assessment problems. One important theme that initially arose in their discussion was the importance of the role of the context in the problem solving process. For example, Mia commented, "The photo album thing threw them off," for which Megan and Mallory nodded their heads in agreement. Specifically they were referring to a multiplicative grouping question in which the pages of a photo album were the groups, and the photos on each page were objects in each group. Apparently, their assigned children were not familiar with the use of albums as a way organizing photos, which in turn made the mathematics problem inaccessible to the children.

In response to Mia's comment on context, Megan added, "For the cupcake [question], [my student] really wanted to do it…" as she began to describe her student's enthusiastic attempt to draw out a picture of boxes of cupcakes.

Immediately after the assessment discussion, the lesson study groups began the lesson planning process for the research lesson to be conducted two days later. In this process, each group planned a research lesson for the three or four children they had interviewed. This second modification involved teaching the research lesson to a small group instead of to the whole classroom, but to a smaller subset of three to six children in that classroom. This constraint was necessary for two reasons. First, given the number of lesson study groups in this university course, there would have been logistical difficulties in trying to find all the classrooms necessary for each lesson study group to have their own elementary school classroom. Second, as novice preservice teachers in the formative stages of learning the mathematical concepts and pedagogy necessary to teach school-aged children, the task of teaching a research lesson to a group of 20 or more school-aged children, as opposed to a small group, could have been overwhelming, and could have drawn their attention away from a focus on their elementary students' learning and the related pedagogy to issues of classroom management (as discussed in the chapter by Fernandez and Zilliox, this volume). Using this modified structure, the lesson study group planned a lesson for their group of children, and one member of the lesson study group was chosen to teach the lesson while the other members observed. During the first lesson planning session, three issues were raised by the preservice teachers. The issues raised were (1) the uncertainty of CGI instruction, (2) the appropriateness of the problem contexts, and (3) the diversity of their children's mathematical content knowledge.

For example, as Megan, Mallory, and Mia began planning their first lesson, for which Megan was the designated teacher, Megan began the discussion by posing a question to the group, "So what's a context we can use? Should we try candy?"

"Like Halloween?" suggested Mia. The general response to the suggestion was tepid, so the group began to brainstorm about other possible contexts as they discussed the different interests their second-grade students had shared on the initial assessment questions.

Amidst the discussion on context, Megan shifted the focus to mathematical content: "What did [your student] say they liked about math?"

Mia began to share with the others ideas based on her elementary students' responses to the assessment. She eventually came to an important point about choosing problems with appropriate number sizes: "For my [two] girls, we need something that is bigger than the number facts that they know, but not too big that they can't solve the [multiplication and division] problem."

As the discussion shifted towards mathematical content, Megan added, "I just don't know how to get [my student] to begin to do multiplication or division, because no matter what, he will try to add the numbers.… He doesn't even try to listen to the context to see what's happening in the problem. It seems like he doesn't have an operational concept of multiplication or division.…" They continued brainstorming. Mia suggested starting first with a multiplicative grouping problem with small numbers.

Mallory added, "Maybe we should have [the students] start easier and get challenging."

Research Lesson

During week 2 of the lesson study project, each lesson study group implemented their research lesson with their small group of children. The lessons were typically 45–50 minutes long. The lesson study groups were usually spread across the classroom and an overflow room, like the school library, to provide the necessary learning space for each group. Typically, the designated teacher sat at a small table or on the floor with the group of students seated around the table or in a circle on the floor. The designated observers were instructed to sit along the perimeter of the learning space. As the teacher conducted the research lesson, the observers were told by their instructor not to interfere with the lesson or interact with the children. In essence, they were learning to observe. This was not an easy task for many of the preservice teachers in that their natural tendency was to assist or teach the struggling learners rather than observing and documenting the children's mathematical behavior.

Debriefing

Immediately following the research lesson, the lesson study groups were given about 20 minutes for the debriefing process. The designated teacher was given the first five minutes to reflect on how she thought the lesson went. The remaining 15 minutes were spent with the observers' reflections. It was stressed by the college instructor that the observers should not critique the designated teachers' instructional decisions, but rather focus on the children's thinking, or how the children seemed to interpret the research lesson tasks. During the actual lesson debriefing time, comments by the professor were limited in that there was not enough time to sit and participate in the debriefing process with each lesson study group. Rather the professor circulated around the classroom to keep the lesson study groups on task while offering short reflective comments on the research lesson when appropriate. Additionally, each group was required to post on an online forum a summary of the issues discussed in the debriefing after each research lesson, but before the next lesson planning session. This provided valuable feedback to the professor about what transpired in each lesson, while providing a means for the professor to also post specific feedback on the discussion board to each lesson study group for consideration in the next planning session.

The Iterative Cycle

During week 3, the lesson study groups returned to their university classroom and were given an hour to revise their research lesson based on their lesson observations, lesson debriefing, and posted online recommendations from the first research lesson. Another preservice teacher took the role of designated teacher for the revised lesson that was to be taught to the same group of children. Teaching a revised lesson to the same group of children by the same lesson study group was a significant modification from what is usually expected in a typical lesson study process. This modification was necessary, since, in this educational context, the preservice teachers did not have their own separate classroom of children with established student-to-teacher relationships. So logistically, in the second and third lessons, the learning group of children remained the same, while the preservice teachers in the lesson study group rotated through the designated teacher position. This gave each preservice student an opportunity to conduct one research lesson and observe two lessons. This also provided a sense of relational continuity for both the university students and their respective elementary students.

For this particular lesson study project, the content area of the second and third lessons remained single-digit multiplication and division. However, the professor suggested to the whole class to change the contexts of the CGI word problems so that the subsequent lessons were not identical to previous lessons, and to adjust the number sizes to better reflect their children's mathematical knowledge. This ensured that the subsequent lessons were sufficiently different to be considered three research lessons, with each lesson going through one lesson study cycle.

One of the benefits of the iterative lesson study cycle observed in this teaching experiment was the development of the college students' depth of understanding of issues discussed in this course. For example, when comparing their planning of the first research lesson to their planning of the third research lesson, the preservice teachers still discussed issues surrounding the nature of CGI instruction, appropriate problem contexts, and their children's mathematical content knowledge. However, as the following discussion indicates, there appeared to be a depth of their understanding and ability to reflect on the lesson planning process in the third lesson planning session that was not as evident in the first lesson planning session.

Planning the Third Lesson

As Megan, Mia, and Mallory began planning the third research lesson, they discussed overall goals for the context and mathematical content. Mallory began the discussion reflecting out loud, "I think that they were interested in the animals. It was something they could all relate to."

As the other two nodded their head in agreement, Mia added, "We want to try to do more base 10 stuff right?" Again, there was a quick consensus.

Megan continued, "[In the previous lessons] if the context was appropriate, they seem to do OK with a multiplication grouping question." As the other two nodded affirmatively, Megan continued to reflect on her specific student, Adam, who seemed unable to grasp the concepts of multiplication. This prompted all three group members to refer back to their teaching notes from the previous lessons to remind themselves what this child did on the math problems.

Mallory continued the discussion, "Well, I had a lot more ideas on the last [lesson] than this [lesson]...I don't know, I think the story works well, but with questions like we had in the middle [of lesson 2], not just the beginning [of lesson 2]." In particular, the three had observed that the multiplication and division problems involving groups of children given at the beginning of lesson 2 did not seem as engaging as problems that involved groups of animals given in the middle of lesson 2. They then began to discuss possible scenarios that involved groups of animals that could be used in their mathematics problems. Mallory then added, "I feel like we're missing something, like we talked about [something in the debriefing of lesson 2] that we wanted to change [in lesson 3]...that we wanted to do different."

Megan responded, "I think we wanted to try another problem with 5s, right? We wanted to work with more 10s, we wanted to try place value…to see what they know about place value."

This gave focus to Mallory, "Okay, let's try a question using place value."

Mia added, "Like try grouping?" As Megan nodded affirmatively, Mia asked, "How would you do grouping…[with place value]?" Megan started to brainstorm some ideas around place value problems as Mia reiterated, "I don't understand how to incorporate place value in word problems…like that's what I'm struggling with."

Megan then suggested, "We could use 10s [in the word problems] and see if they count by 10s, or umm…or whatever…after they get their answer you could ask them questions to see if they understand place value. If their answer is 100 or something, see if they know each place. So, like 120, see if they know what the two means…you know what I mean?"

Up to this point, most of the lesson planning focused on the role of context and mathematical content in the upcoming lesson. In particular, it was recognized by the group that certain contexts seemed to elicit more participation from the elementary students. In this case, the students seemed to respond better to problems involving groups of animals rather than groups of people. The preservice teachers also discussed the relationship between context and mathematical content as they struggled to write word problems that directly reflected concepts of place value. This eventually led to a discussion involving children's mathematical knowledge, the importance of a meaningful understanding of the representational tools used to model mathematical thinking, and the continued tension of using child-centered instructional strategies instead of teacher-directed instruction.

Mia suggested a possible question that involved five groups of ten. Rather than discussing the suggestion, Megan pursued another line of thought unrelated to place value: "I'm trying to think what each kid knows individually…like how they solved each problem last time [in lesson 2]." In particular, she came back to her elementary student, Adam, for which the group discussed his lack of motivation in solving the problems in the previous lessons. However, Megan did share her observation that Adam, when using unifix blocks, was better able to model the mathematics in the questions he attempted.

This idea led Malory to share, "Sometimes it's hard, like when [Adam] was doing the modeling, and [the teacher] asked, 'So what does this [block] mean?' … And I think it's easy for kids to be, like, 'What do you mean what does this mean?'" Megan and Mia nodded their heads in agreement.

Megan added, "Yeah, like trying to get them to understand the connection between…this block actually stands for something [mathematically]…. And it's hard to get them to understand what you're talking about without telling them."

Mia continued, "That's what I was trying to say [when I was teaching lesson 2 by asking the kids], 'O.K. is this [block] a group of people? Is this [block] one person? What is this [block],' to try…and then they would tell me, 'Oh, it's a group of people.'"

The issues described in the previous excerpt continued to frame the remainder of the lesson planning discussion. Afterwards, the lesson was taught and debriefed as expected in the iterative cycle of the lesson study process. The benefit of this iterative cycle was an increased depth of understanding of the complexities in teaching mathematics. While these complexities were not resolved through this process, an opportunity to re-engage these issues through group discussion and teaching experimentation was provided through this iterative process.

Reflections and Conclusion

This chapter provides a view of lesson study in the context of preservice elementary education teachers. While the broader goals (site-based, practice-oriented, collaborative, and centered on school aged children's learning) of lesson study remain the same across both contexts, the use of lesson study in early field-based experiences have the added goal of providing a cohesive structure to meaningfully frame the preservice teachers' learning experience. In the case of this particular college course, the modified lesson study model (Table 2), while abbreviated in nature, provided a collaborative environment for the students to engage in and make sense of their first teaching experiences in mathematics as they collectively tried to make sense of their designed mathematical tasks and how the elementary students learned through problem solving.

Furthermore, professional growth was exhibited through the preservice teachers' improved ability to design meaningful problems and analyze their student's mathematical understanding. However, given the modifications to the lesson study process as described in this paper, there are some limitations to consider. First, in this modified lesson study model, the learning goals were assigned to the preservice teachers. This is a limitation in that one purpose of lesson study is to create professional responsibility with teachers as the teachers themselves formulate the long-term goals of student learning and development. This responsibility was not fully achieved as the preservice teachers had allowed to determine the learning goals of

Table 2 A comparison of this preservice teacher lesson study model with a more typical inservice lesson study model

Lesson study goals	Typical study with inservice teachers	Modifications of lesson study with preservice teachers
1. Formulate long term goals of student learning and development	Inservice teachers may have more latitude in determining the learning goals for their research lessons	The learning goals were given to the preservice teachers by the university instructor
2. Plan, conduct, and observe a "research lesson"	Research lesson is conducted in a whole classroom of school-aged children	Research lesson is conducted to a smaller group of three to six children
3. Discuss and revise the lesson and approach to instruction based on these observations	Revised lesson is taught to a different classroom every time	Revised lesson is taught to the same group of children by the same lesson study group

their lesson study. Second, by teaching the research lesson to a smaller group of children, the preservice teachers are not fully exposed to the complexities of teaching a whole class, and in particular, how their research lesson would be conducted in a whole class. A third limitation is that by teaching a revised lesson to the same group of children, the preservice teachers did not get the diverse feedback a different group of children would provide. None the less, the use of this modified lesson study provided a viable professional development framework to help the preservice teachers develop a deeper conceptual understanding of mathematics, expose them to research based pedagogy, and provide them with a meaningful field-based experience to investigate children's understanding of mathematics.

References

Carpenter, T. P., Fennema, E., Loef-Franke, M., & Empson, S. B. (1999). *Children's mathematics: Cognitively guided instruction.* Portsmouth: Heinemann.

Grossman, P., Wineburg, S., & Woolworth, S. (2001). Toward a theory of teacher community. *Teachers College Record, 103*(6), 942–1012.

Stafford-Plummer, J. (2002). *An analysis of the influence of lesson study on preservice secondary mathematics teachers' view of self-as-mathematics expert.* Unpublished master's thesis, Brigham Young University, Orem, Utah, USA.

Takahashi, A., & Yoshida, M. (2004). Ideas for establishing lesson study communities. *Teaching Children Mathematics, 10*(9), 436–443.

Zeichner, K. M. (1981). Reflective teaching and field-based experience in teacher education. *Interchange, 12*(4), 1–22.

Response to Part II: Emerging Issues from Lesson Study Approaches in Prospective Mathematics Teacher Education

Despina Potari

In this chapter, I will draw on the three chapters by Fernandez and Zilliox, Murata and Pothen, and Yu. I will identify emerging issues across the three chapters that seem to be central to mathematics teacher education. In particular, I will discuss the way lesson study has been implemented in the different contexts; the prospective teachers' learning about teaching mathematics; and the main factors that framed learning. Finally, I will attempt to offer some ideas that could be considered in lesson study research.

Lesson Study Approaches

The three chapters in this part implemented lesson study within mathematics methods courses that were linked to field-based experiences in a teacher education program. Two of the cases describe prospective elementary-school teachers. The third is about prospective secondary-school mathematics teachers. All the cases follow the typical structure of lesson study but they have variations that are a result of the different theoretical perspectives adopted, the particular characteristics of the prospective teachers, and the structure of the teacher education program and field-experience. In Table 1 the particulars of each case are summarized. The primary difference between the studies is the way the authors/teacher educators integrate theory and practice. In preservice teacher education, this relationship is very important and very difficult to achieve as prospective teachers have limited classroom experience and frequently hold traditional beliefs of mathematics and mathematics teaching. As a result, the meaning that they attribute to theory remains at a rather superficial level (e.g. Ambrose 2004; Stuart and Thurlow 2000). The dialogic cycle of knowledge creation proposed by Ruthven (2002) where scholarly (academic) knowledge is interrelated to craft knowledge (knowledge emerging from practice) through interaction between research and teaching is built in the lesson study approaches of

D. Potari (✉)
University of Athens, Athens, Greece
e-mail: dpotari@math.uoa.gr

L. C. Hart et al. (eds.), *Lesson Study Research and Practice in Mathematics Education*, DOI 10.1007/978-90-481-9941-9_10, © Springer Science+Business Media B.V. 2011

Table 1 Main characteristics of the lesson study approaches

Chapters	Lesson conducted	Particular focus
Fernandez and Zillox	In classes of different grades of elementary school	• Developing theoretical knowledge before lesson study (research findings, theories, teaching materials, skills) • Research lessons involved hands-on experiences • Production or written reports and oral presentations
	Microteaching in small groups of six prospective secondary mathematics teachers	• Developing theoretical knowledge before lesson study (research findings, theories, teaching materials, skills) • The use of a survey to identify content areas where prospective teachers faced difficulties • The instructor as a resource and provider of feedback • Production or written reports
Murata and Pothen	In classes of different grades of elementary school	• Readings about lesson study, watching relevant videos, research-related readings • A variety of assignments (e.g. teacher interview data, student assessment data, portfolio, reflection paper) • The written guidelines
Yu	In small groups of second and third grade children (the same group in the revised lesson)	• Developing theoretical knowledge before lesson study (Cognitive Guided Instruction (CGI) based lessons) • Assessment interview • Focus on the role of context in mathematical learning

these three cases in different ways. So we discern that all the programs support prospective teachers' development of knowledge related to research and theory of mathematics education.

This support, however, had different forms in each program. For example, we see different levels of prospective teachers' involvement in building theoretical knowledge in mathematics education, varying from attending course presentations to personal readings. We also see different ways of connecting research to actual practice through a number of activities, such as designing and analyzing student assessment interviews based on research findings, modifying research tasks (e.g. CGI word problems), interpreting students' mistakes, and justifying teaching decisions and expectations in the light of existing research. In all cases, the prospective teachers received feedback on their debriefing and oral presentations, either from their instructors, the cooperating classroom teachers, or their peers. In the case of Murata and Pothen, teacher educators supported prospective teachers' lesson planning and reflection through written guidelines and the structuring of different assignments.

Another difference among the proposed approaches is the context in which the lesson was conducted. In two cases, the research lessons were implemented in classes where the prospective teachers were doing their teaching practice and had previous experience with the context by observing and teaching the classes. To avoid management problems, Yu's prospective teachers worked with small groups of children. In the case of the prospective secondary teachers, the lesson study involved teaching mathematics to their peers. In all cases, emphasis was made on the development of knowledge of the prospective teachers by elaborating on their conceptual understanding of certain mathematical ideas and by investigating children's mathematical thinking. The development of this knowledge in each case is discussed below.

Prospective Teachers' Learning

In all three chapters, the authors indicate development of prospective teachers' learning. Murata and Pothen discuss the ways various lesson study activities facilitate learning. They claim that the preservice teachers come to understand the importance of student learning in the lesson study process, and through this exploration they enhance their understanding of mathematics content. Moreover, they develop management skills for designing and evaluating lessons as well as interacting effectively with the children. They also learn how to collaborate with both experienced and inexperienced teachers to develop a "shared responsibility for all students." Finally, the prospective teachers make connections between teaching, student learning and theory/research.

In Yu's chapter, evidence is given that prospective teachers make connections between teaching, learning, and research through the analysis of the issues that one group of three prospective teachers raised. They questioned the role of context in mathematics teaching and learning, the nature of CGI instruction, and the children's mathematical knowledge. By planning, analyzing, and revising their teaching, the three teachers developed deeper understanding of the complexity of mathematics teaching and learning through the identification of interconnections between aspects of teaching and learning. In particular, they related mathematics content, context, and learning in task design and assessment, developing their mathematics knowledge for teaching (Ball et al. 2008).

In the chapter by Fernandez and Zilliox, learning was investigated through Bransford et al. (2000) features of learner-centeredness, knowledge-centeredness, assessment-centeredness, and community-centeredness. In terms of learner-centeredness, the lesson study approach helped prospective teachers reconsider and modify ideas about teaching mathematics and move towards reform-oriented mathematics teaching. The prospective teachers also developed deeper understanding of mathematics and mathematics teaching through their engagement in "sense-making, metacognitive processing and reflection." For assessment-centeredness, the authors noted that the support from peers and instructors seemed to help prospective teachers pay more attention to students' thinking and less to classroom management. Finally, the community of beginning teachers who collaborated in the lesson study context learned how to work with others, share experiences, and form an identity of belonging to the wider community of beginning teachers.

Learning reported in the three chapters refers to the development of content knowledge, knowledge of students' mathematical thinking, and pedagogy, as well as to the ability to cooperate with others and develop a sense of belonging to a community. These aspects of learning were identified in inservice teacher education in the context of lesson study in the USA by Lewis et al. (2009). In the next section, I elaborate on the main characteristics of the lesson study approaches that framed prospective teachers' learning.

Factors that Framed Learning

Because of the integrated nature of lesson study, it is difficult to isolate specific factors that frame prospective teachers' learning. In the previous section, I addressed issues that according to the authors appear to have an effect on prospective teachers' learning, such as the connection of theory to practice, the lesson study cycles, the group collaboration, the feedback provided by teacher educators, cooperating teachers, and peers. In this section, I will elaborate on two of these issues that appear to be crucial in the professional development of beginning teachers: the role of the teacher educator and group collaboration.

The Role of the Teacher Educator

Research on the role and the development of the teacher educator in teacher change has received interest over the last decade (see Zaslavsky and Leikin 2004; Garcia et al. 2006). In the three chapters, the teacher educator is a facilitator of prospective teachers' learning through different degrees of guidance and support. In the chapter by Fernandez and Zilliox, the authors describe the positive effect of the instructor on the prospective secondary teachers contrasted to the limited guidance given to the elementary teachers. In the former group, the instructor was part of the lesson analysis within the lesson study cycles, collaborating with the prospective teacher and posing questions while the elementary case feedback was provided only to oral reports. The "knowledgeable other" played an important role in helping prospective secondary teachers to develop their understanding of mathematics teaching and learning. Without this guidance, the prospective elementary teachers focused on classroom management issues and not on student mathematical thinking and its development. In the case of the lesson study approach reported by Murata and Pothen, the teacher educator guided prospective teachers through a number of assignments. The teacher educator helped prospective teachers notice students' mathematical activity. She supported them emotionally and helped them manage the various lesson study activities while encouraging them to reflect on their actions. In Yu's chapter, the teacher educator supported learning by providing the CGI learning goals that the prospective teachers were familiar with in advance of the lesson study cycle.

Lesson study feedback was in the form of reflective comments in person or in an online forum.

In general, the teacher educator played a crucial role in the learning that occurred during the lesson study process in all three cases. This is not always the case for inservice teachers. Learning can also be discussed in relation to what the teacher educators are learning about their students and about their role in the process (see the chapter by Watanabe on *learning stance*, this volume). This raises important questions. Are the teacher educators members of the lesson study community? Are they coresearchers? What distinguished the role of the expert in the lesson study context to that of a collaborator in a community of inquiry of mathematics teaching described by Jaworski (2003) in the case of inservice teachers?

The Group Collaboration

In lesson study, group collaboration is key. In all three chapters group collaboration was highly valued by the prospective teachers as they had the opportunity to share ideas, see teaching and learning through different lenses, develop a sense of belonging, and in most cases develop their understanding of mathematics teaching toward reform-oriented practices. In the group interactions, questions posed both by participants or by the knowledgeable advisor encouraged reflection and led to deeper understanding. The overall experience of collaboration provided a model for future interactions and collaborations with colleagues. This too raises important questions about the ways this collaboration develops. How do the groups reach a level of defining common goals? What type of interactions, negotiations, and contradictions emerge, and how do they influence the process of inquiry on mathematics teaching? What are the similarities and differences between the collaboration of prospective teachers and inservice teachers? In what ways can existing research frameworks for studying teacher collaboration help us to investigate the form and the content of collaboration?

Concluding Remarks

The three chapters in this part demonstrate that lesson study is an effective way for prospective teachers to build relations between theory and practice, develop their mathematics knowledge for teaching and the knowledge of students thinking, and form a teacher identity towards an inquiry approach to teaching through collaboration with peers and knowledgeable advisors. In this report, I attempted to identify issues that emerged in the different approaches to lesson study at the preservice level and posed some questions for further consideration at the research and practice level. Another direction forward would be to investigate lesson study at the preservice level beyond the US experience presented here by taking into account the existing culture, traditions, and contexts of mathematics teacher education and research on mathematics teacher education in other countries. Moreover, to compare lesson study to existing international theory and research on mathematics teacher education would improve our understanding of lesson study and provide us with frameworks to investigate further its role and importance for preservice mathematics education at an international level.

References

Ambrose, R. (2004). Initiating change in prospective elementary school teachers' orientations to mathematics teaching by building on beliefs. *Journal of Mathematics Teacher Education, 7*, 91–119.

Ball, D. L., Thames, M. H., & Phelps, G. (2008). Content knowledge for teaching: What makes it special? *Journal of Teacher Education, 59*(5), 389–407.

Bransford, J., Brown, A. L., Cocking, R. R., & CBSSE. (2000). *How people learn: Brain, mind experience, and school.* Washington: National Academy Press.

Garcia, M., Sanchez, V., Escudero, I., & Llinares, S. (2006). The dialectic relationship between research and practice in mathematics teacher education. *Journal of Mathematics Teacher Education, 9*, 109–128.

Jaworski, B. (2003). Research practice into/influencing mathematics teaching and learning development: Towards a theoretical framework based on co-learning partnerships. *Educational Studies in Mathematics, 54*, 249–282.

Lewis, C. C., Perry, R. R., & Hurd, J. (2009). Improving mathematics instruction through lesson study: A theoretical model and North American case. *Journal of Mathematics Teacher Education, 12*, 285–304.

Ruthven, K. (2002). Linking researching with teaching: Towards synergy of scholarly and craft Knowledge. In L. D. English (Ed.), *Handbook of International Research in Mathematics Education* (pp. 581–598). London: Lawrence Erlbaum.

Stuart, C., & Thurlow, D. (2000). Making it their own: Preservice teachers' experiences, beliefs, and classroom practices. *Journal of Teacher Education, 51*(2), 113–121.

Zaslavsky, O., & Leikin, R. (2004). Professional development of mathematics teacher educators: Growth through practice. *Journal of Mathematics Teacher Education, 7*, 5–32.

Part III
Challenges and Promises of Unchartered Water: Lesson Study and Institutes of Higher Education

Lesson Study as a Tool for Developing Teachers' Close Attention to Students' Mathematical Thinking

Alice S. Alston, Lou Pedrick, Kimberley P. Morris and Roya Basu

Introduction

Does lesson study "work" as an effective tool for professional development? What can be documented to show learning that might be of help to others? These questions guided classroom-based professional development and collaborative research among teachers and university educators within a long-term partnership between an urban school district in central New Jersey and Rutgers University. Overall, the partnership aims to support the district as it addresses the challenge of increasing mathematics achievement among all students. The goal of this particular project has been to strengthen middle-grades teachers' ability to attend closely to the mathematical activity of students and to develop effective pedagogical strategies while simultaneously engaging as learners of the mathematics that they are expected to teach. This chapter, jointly prepared by university educators at the Robert B. Davis Institute for Learning (RBDIL) and two teacher researcher/participants in the lesson study, describes findings from one complete cycle of lesson study activities in an emerging model of school-based professional development using a modified Japanese lesson study within the structure of an off-campus graduate course.

In this two-tiered study, the first tier includes two teachers' analyses of their students' mathematical activity and reflections about the implications of their findings for their teaching. The second tier assumes the perspective of the university researchers, for whom the unit of analysis was the group of teachers as they experienced and reflected about the lesson study. Specific to this analysis was the identification of

This research and the professional development program that provided its context was supported by a Standards Implementation Grant from the New Jersey Statewide Systemic Initiative. The views expressed in this paper are not necessarily those of the funders.

Teacher participants in the lesson study include Ranilla Ahmed, Roya Basu, Robert Birkitt, Ronald Foley, Rosalyn Gallmon, Fred Harris, Laurey Leivonen, Victor Liu, Kim Morris, and Sarah Pretty.

A. S. Alston (✉)
Rutgers University, New Brunswick, NJ, USA
e-mail: alston@rci.rutgers.edu

L. C. Hart et al. (eds.), *Lesson Study Research and Practice in Mathematics Education*,
DOI 10.1007/978-90-481-9941-9_11, © Springer Science+Business Media B.V. 2011

any evidence of the teachers' understanding of mathematics, their attention to students' mathematical behavior, and inferences that the teachers drew from this experience for their own instruction. Data was collected during one lesson study cycle, including ten weekly preparatory sessions, six classroom implementations in grades 5–8, subsequent debriefing sessions and written analyses and reflections.

Research Questions

The following questions guided the study:

1. What ideas, actions, explanations, and justifications offered by the students during the implementations of the research lessons do the participating teachers identify as evidence of understanding or misunderstanding of the mathematical ideas in the lesson?
2. What evidence is there that teachers have deepened their attention to and understanding of students' thinking about the mathematical ideas that are being addressed in the research lesson(s)?
3. What evidence is there that participating teachers are becoming more aware of their own pedagogical decisions and the effect that their interventions have on their students' learning?
4. What evidence is there that the use of videotaping enhances the teachers' awareness?

Theoretical Framework

Research in mathematics education over the past two decades was strongly influenced by the publication of the standards documents from the National Council of Teachers of Mathematics (1989, 2000). Specifically, the implicit expectation that students should be able to successfully construct their own understanding of basic mathematical ideas and use this knowledge to solve challenging and unfamiliar problems has profound implications for mathematics instruction. Implementing these standards has been difficult, raising questions about instruction and preparation for instruction for practitioners, teacher educators, and mathematics education researchers at every level. In reporting the results of the Third International Mathematics and Science Study (TIMSS), Stigler and Hiebert (1999) concluded that then current teaching practices and content knowledge among teachers in the United States needed to change if students were to be adequately prepared to meet the challenges of the twenty-first century, and that positive change presupposed a different approach to preparing and supporting teachers. Ball and Bass (2003) discuss the importance for students of *making mathematics reasonable*, including their active engagement in solving and justifying challenging problems. This is consistent with the work of Maher et al. (2007) with its close attention to students' mathematical reasoning and the value of teachers building instruction based on the representations and models that their students construct (Maher 2008; Laffey 2008).

Lesson study as a tool for professional development is among the recommendations for supporting teacher change outlined in *The Teaching Gap* (Stigler and Hiebert 1999). Based on the results of their study of Japanese instruction and professional development, lesson study and its general benefits to instruction were found to be consistent with research in the United States indicating that meaningful and lasting teacher growth occurs over time as a result of on-going professional support that is directly related to the teachers' classroom needs (Loucks-Horsley et al. 1998). Providing evidence of the value of such support depends on careful analyses of actual professional development implementations that focus on qualitative characteristics of both learning and teaching (Hill et al. 2005). Making a persuasive case for the importance of this form of professional development becomes critical, if it is to remain a tool for supporting instruction that is truly standards-based and encourages the development of environments in which students have the opportunity to investigate challenging mathematical problems, select strategies that work for them, and build meaningful representations of their solutions to share and defend with their peers.

One of the possible structures available to lesson study is the use of video as a tool to enhance practitioners' natural ability to reflect on their practice. Although widely used in educational research, video has yet to achieve a mainstream position in everyday classroom practice. Current research indicates that the use of video as a tool contributes to the goal of bringing educational practices to a level where teachers are indeed masters of their craft with the goal of empowering students toward success in mathematics (Maher 2008). Derry (2007) points out that use of video has potential to reduce biased interpretations since the content could be available for unlimited review. Studying videos in groups provides a context in which to share multiple points of view. The ability to carefully analyze the classroom video affords the teachers an opportunity to observe and study their students' mathematical activity in a different light (Goldman 2007).

Background and Methodology

Lesson study as a process for professional development has evolved as an important activity for teachers within partnerships between the Robert B. Davis Institute for Learning (RBDIL) and various school districts in central New Jersey over the past decade. Central characteristics of the lesson study approach as described by Lewis (2002) and by Murata in this volume are consistent with the RBDIL philosophy of long-term, school-based professional development and research, specifically those activities that focus on teachers' deepening knowledge of the mathematics that they are expected to teach within a context of close attention to their students' mathematical thinking. This focus has proven particularly effective when enhanced by concurrent study of videotapes of children engaged in thoughtful mathematical activity (Goldman 2007, Maher 2008). Project reports from RBDIL (Berensen 2004, 2005) and analysis by Alston et al. (2005) provide documentation of the effectiveness of lesson study within RBDIL professional development partnerships. The example of collaborative research embedded within an intensive professional development program described in the present study illustrates lesson study as part of an off-campus graduate course in mathematics education situated in a partner school district.

Ten teachers, representing grades 5 through high school, were participants in a graduate course offered in a middle school classroom in central New Jersey through the Department of Continuing Education of the Graduate School of Education of Rutgers University. Two of the authors, a faculty member of the university and a doctoral student, formerly a middle school teacher in the district, co-taught the course. Five of the teachers were currently teaching in the district. Each of the five had participated in earlier courses and professional development activities with RBDIL, and was knowledgeable about lesson study. The remaining five teachers were teaching in neighboring districts.[1] Based on analysis of the previous year's activities that focused on problems involving combinatorics and rational numbers (Alston et al. 2007) and discussion of the specific needs of their students, the teachers recommended that topics foundational to algebra should be the mathematical focus for the lesson study reported here. The course met on Thursday afternoons, January through May, from about 3:30 until 6:00 to study the development of algebraic ideas from 5th-grade through 9th-grade Algebra I, solve related problems together, read and discuss relevant literature, identify specific needs for their students and develop a common outline of activities for research lessons that could be modified appropriately for each of their classrooms. From the first session, the participants were a part of Sakai (a web-based network at Rutgers University) through which there were regular communications about course activities, assigned readings and mathematical activities, and a discussion panel that allowed for interaction about various issues between the weekly sessions. Early sessions included an investigation of the development of algebraic ideas in the curricular resources and assessments used by the district: Connected Mathematics Project (CMP) in middle grades and Everyday Mathematics in the elementary schools. Frequent reference was made to state and national standards and to the expectations defined by the state assessments. The Algebra I textbooks used by the district were consulted for evidence about how particular ideas would be presented in the traditional course that is a requirement for high school graduation.

The final project for the graduate course was a series of six research lesson implementations and debriefings. Six of the ten participating teachers, representing grades 5, 6, and 8, offered to facilitate the public lessons. A videographer using a single roving camera videotaped the six lessons and subsequent debriefings. The remaining four teachers, representing grades 6, 7, and 9, were expected to implement

[1] Two of the teachers continued their analyses beyond the scope of the course, one for a master's thesis and the other as the basis for a doctoral dissertation proposal. Their analyses and collaboration in developing this chapter led to their role as coauthors.

Table 1 Schedule of six lesson study implementations

Day	Morning		Afternoon	
1	Grade 5 lesson	Debriefing	Grade 6 lesson	Debriefing
2	Grade 8 lesson		Debriefing Day 1 and 2	
3	Grade 5 lesson	Debriefing	Grade 6 lesson	Debriefing
4	Grade 8 lesson		Debriefing Day 3 and 4	

and report on some modification of the research lesson with their students. All were expected to participate as observers in at least one day of the public lessons and subsequent debriefings. See Table 1 for a summary of the public lesson implementations.

Using an observation protocol modified from Lewis (2002), observers at the lesson implementations were expected to complete and submit careful observation notes about the children's mathematical activity during the lesson. Observers were not to interact with the children and were to scatter themselves among the students so there could be comprehensive reporting of what the students did mathematically. Each of the teachers in the graduate course was present as an observer for the first two days of implementations and debriefings. The three subsequent lessons (Days 3 and 4 on the chart) took place two weeks later. All of the implementations were open to interested teachers, administrators, and mathematics coaches in the district. Several attended each. For Days 3 and 4 the teachers found it difficult to be released to observe. Consequently, for the last three lessons, only a few of the other participating teachers were observers. The two course instructors were observers at each of the implementations and facilitated the debriefing discussions. Including the instructors, there were at least five observers at each of the lessons and as many as fifteen observers for the first three implementations. Each of the facilitating teachers was provided a copy of the videotape for his or her session to view and refer to in writing the final reflective paper.

Data for the following description and analysis of results for the lesson study include observation notes of the instructors, notes recorded by the teachers on the on-line Sakai class discussion forum, worksheets for the activities, videotapes of the six public lesson implementations and debriefings, students' work from the implementations, the teachers' written reflections after the implementations, and the in-depth-written analyses of the two coauthor teachers, including their transcripts of selected episodes from the videotapes of the research lessons and debriefings that they considered critical examples of their own actions and of their students' thinking.

Selected Results

The Planning/Investigation Phase

For each mathematical concept considered in the course, teachers first engaged in problem activities themselves, sharing their representations and solutions. These

Fig. 1 Shorts and glasses with price

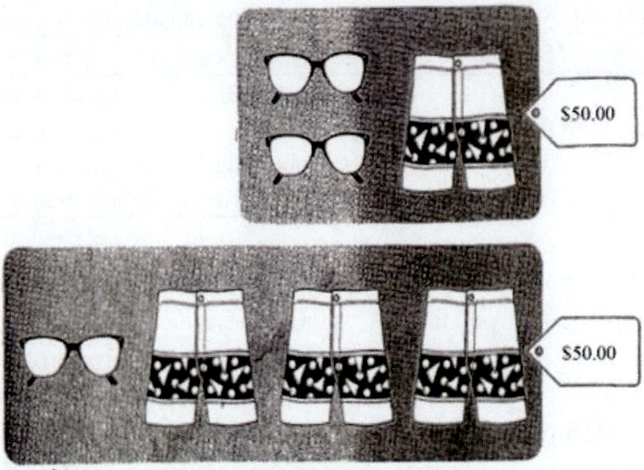

1. Without knowing the prices of a pair of glasses or a pair of shorts, you can determine which item is more expensive. Explain how.

2. How many pairs of shorts you buy for $50?

3. What is the price of one pair of glasses? Explain your reasoning.

sessions, facilitated by the mathematics educator instructor as the "outside expert," were intended to model environments similar to those appropriate for implementing the problem activities with students, and included discussion about pedagogical decisions, teacher questions that might foster the building of ideas, and expectations that the teachers had for their students. The teachers regularly took these problems back to their classrooms and implemented them with their students. Observation notes and samples of student work were brought to subsequent classes for discussion. For example, the Shorts and Glasses problem in Fig. 1 (Meyer and Pligge 1998) is a task the teachers first solved themselves during a class session and subsequently gave to their students. At a later class, several of the teachers reported how their students approached the problem. Sarah reported:

> I had a chance to try the Price Combinations problems (see Fig. 1).... They were all over the place in terms of responses; some pulling other information into the problem like the glasses cost more because they have an eyeglass cleaner with them!... They seem to understand the concept of equality although one student's solution to..."What is the price of one pair of glasses?" was the following: "$20.00 because 20 + 20 = 40 + 10 = 50 and 20 + 10 + 10 + 10 = 50." The student's use of the equal sign as shown above is not a true statement. I think this is something we should be "sticklers" about so that the kids understand equality better.

Laurey, a middle school special needs teacher, posted this comment in response to an on-line question from one of the instructors:

> I also gave them the problem with the glasses and the shorts. They needed time. However they were able to determine that the glasses were more expensive than the shorts. They realized that both pictures were worth the same amount and one picture only had 3 versus 4 items. Knowing that information they determined which was more expensive and then used trial and error to calculate the answer.

Roya, an 8th-grade algebra teacher, commented that her experience was similar to that described by Laurey. She reported that when presented the problem, her students simply selected trial prices for each article and then adjusted them to meet the price constraint of $50. She agreed that "guess and check" was a valid strategy but had as a goal for her 8th-grade students to develop other more "algebraic" strategies as well.

The Shorts and Glasses problem served as a basis for an on-going discussion on equivalence throughout the development of the research lessons. The two 5th-grade teachers in the group reported that a major emphasis in the algebra strand of Everyday Mathematics at grade 5 was equivalence, using an actual balance scale and then drawings of scales to explore the idea. They noted that the children were just beginning to build ideas about the algebraic meaning of "equations" as balanced, equivalent expressions. Noticing that there seemed to be an abrupt shift in the text materials from the experiential use of physical objects to the use of symbolic representation and subsequent manipulation of symbols, the group decided to look for resources that might address this gap in the curricular development.

In another session, the middle school teachers shared their experiences using the first algebra units in CMP. The teachers made two observations. First, they noted that these units do not occur during the 6th-grade year, but rather in grade 7. This resulted in a full academic year of mathematics in which the algebraic ideas introduced in 5th grade were only tangentially addressed. There appeared to be little, if any, development of ideas involving algebraic expressions and equations except for applications of formulas involving area and perimeter. Secondly, in looking closely at the 7th-grade units, the teachers observed that the primary algebraic ideas involved functions and relations, coordinate graphing, and the extension of the number system to include operations with negative as well as positive numbers. There was very little direct emphasis on equivalence and equations.

The discussion expanded further with a reading from *Thinking Mathematically: Integrating Arithmetic and Algebra in Elementary School* (Carpenter et al. 2003) which reported on research about children's understanding of the meaning of the equal sign. In the reading children were given the equation: 8 + 4 = ___ + 5, and asked to place a number in the equation that would make it a true statement. In their analysis, the authors found that a significant number of students placed the number 12 into the blank. The teachers in our graduate course adamantly expressed their belief that their students would not repeat this incorrect notion. They agreed to try it during the next week. All ten reported their surprise when many students did indeed believe 12 to be the appropriate solution.

Roya posted the following in the on-line forum:

> After reading Carpenter's article…I wondered if students in higher grades would make similar mistakes…. So I decided to make a worksheet and gave it to 50 of my students of varying abilities. Two of the questions were similar to those mentioned.
>
> 1. 12 + 6 = ___ + 2.
> 2. 18 − 8 = ___ + 1.
>
> 18% of the 8th graders got these wrong!

As a result of experience with problems such as the Shorts and Glasses problem and the Fill in the Blank problem as well as careful examination of the curricula across the grades, the teachers decided to focus on concepts involving equivalence for their research lessons. They examined various standards-based resources for ideas. The 6th-grade unit from Mathematics in Context (MIC), *Comparing Quantities* (Meyer and Pligge 1998), from which the Shorts and Glasses problem was taken, includes equations and inequalities as pictorial "balanced expressions." The unit presents systems of linear equations based in real-world situations that encourage a variety of representations, including subsequent pictures, tables, graphs, and symbolic equations for finding solutions for the systems of two and three equations. These problems were appealing to the teachers because they did not impose particular symbolic structures, but encouraged the use of strategies and representations that were constructed by and, hopefully, meaningful to the students. After working through the problems themselves and comparing and justifying solutions, which proved to be no trivial exercise, the teachers divided into two groups, one for grades 5 and 6 and the second for grades 7–9. Using the problems from the MIC unit, the teachers developed lesson plans based on modifications of several of the problems. The group followed a lesson study structure (Lewis 2002), including defining specific goals for students, predicting student responses, suggesting possible teacher questions and probes, and thinking about ways for students to share their solutions. A series of problems were selected, including two that used balance scales "weighing" various fruits and vegetables (Fig. 2), an inequality problem based on a "Tug

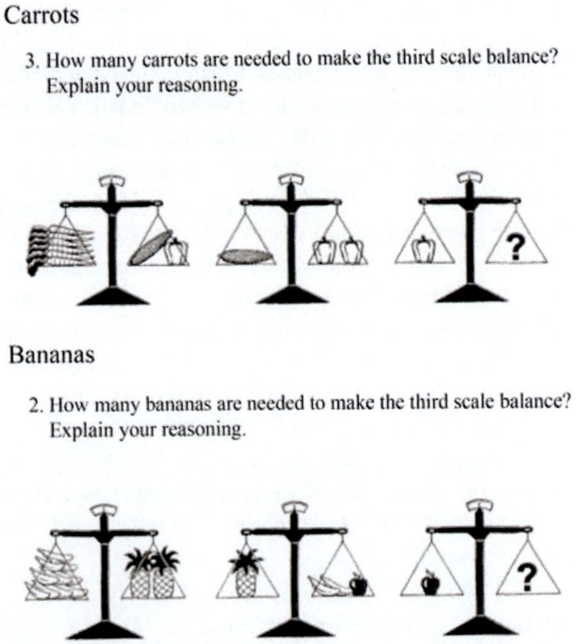

Fig. 2 Carrots and bananas

Fig. 3 Tug of war

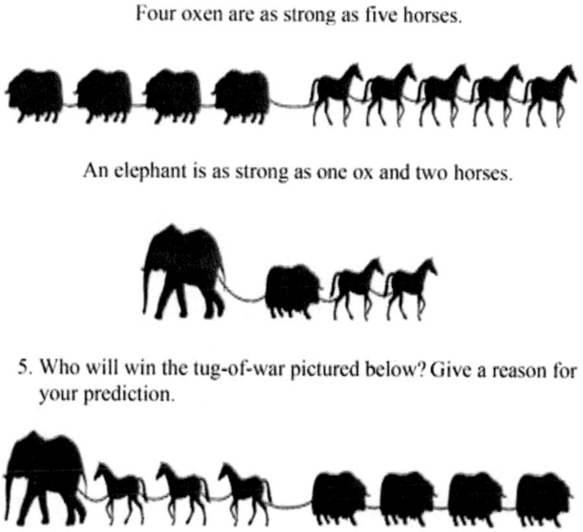

of War" (Fig. 3), and several additional tasks similar to the "Shorts and Glasses" problem discussed earlier.

The Research Lessons

For the first lesson implementation, the facilitating 5th-grade teacher, Sarah, chose to retain the $50 price in the Shorts and Glasses problem. During the debriefing, the teachers noticed that as in Roya's experience, the children approached the problems by substituting numbers rather than analyzing the pictures and recognizing that the value of the sunglasses was double that of the shorts. With the total cost given, the students readily noticed that $10 for shorts and $20 for sunglasses solved the problem.

In the debriefing discussion immediately following, the teachers discussed various aspects of the students' thinking. In particular, they concluded that, because a "guess and check" substitution was relatively easy, there had been no recognition of the two-to-one relationship between the glasses and the shorts. The group decided that for the afternoon 6th-grade class the price would be omitted from the problem statement. Fred, the facilitating teacher for the 6th-grade lesson, proposed to present only the pictures to the students using an overhead transparency with the accompanying statement that the two sets of objects were the same price.

In the debriefing session following the afternoon session, the teachers noted that a variety of strategies had been adopted. Some students did assign a dollar number to each of the items but the values varied among the solutions of different student groups. Because these numbers varied, a number of the students recognized the "two-to-one" relationship between the values of the items. Other students were observed representing the items symbolically and manipulating the equations in order to find an answer to the problem. Several of the teachers noticed that some students did respond to the problem with comments about the context, stating their belief that either the sunglasses or the shorts should certainly be more valuable for nonmathematical reasons.

The third implementation took place the following morning in an 8th-grade class. After observing and discussing the two preceding implementations during the Day 1 debriefings, Kim, the facilitating teacher, decided to modify her proposed plan. She originally planned to begin with the Tug of War problem (Fig. 3) and then to proceed to another task that would demand the construction and analysis of three different equations. On the basis of her observations in the earlier classes and comments made by the observing and facilitating teachers, she proposed to the group that she would like to change her lesson plan and precede the Tug of War task with the Shorts and Glasses problem followed by the two problems involving balance scales (Fig. 2 above) even though previously she thought these problems would not be sufficiently challenging for her students. In her later analysis, Kim notes (Morris 2009, p. 7):

> Fred…. Noted challenges with solving for equality with two variables. While solving for the Shorts and Glasses task, a few students meaninglessly placed value to the items without considering the relationship…. As a result, I revised my lesson plan…to use the activities from the elementary lesson study. I felt it was vital for students to explore solving for equalities and equivalents prior to solving inequalities.

During the first half hour of the lesson, the students were extremely engaged as they built solutions to the three problems and explained their solutions before proceeding to Tug of War (Fig. 3), which involved an inequality and proved to be considerably more difficult.

Sarah, in her written reflection following Kim's implementation, noted the students' ability to substitute values derived from the balance scales.

> The students saw the equivalences from one scale to the next and actually discussed the equivalences in terms of pounds as their unit of measure. This was a significant point to observe because in our teacher discussions we made note of the nuance regarding language used to compare the items on the pan balance. It was important to note that, for example, an ear of corn is not equal to a pepper but that the weight of an ear of corn is equivalent to the weight of a pepper.

After studying the videotape of her lesson implementation session, Kim selected and commented about the following interaction with one of her students as a particularly important example of algebraic reasoning (ibid, p. 9).

Tyrreal's Solution to the Scale Balance Task

> Tyreal: If each one (carrot) equals one fifty, two of them will be 3. Then 3, 6, 9 (counts the pairs of carrots on the first balance)—then this one (the ear of corn) will have to be 6 and this one (the green pepper) will have to be 3. And if this (corn) is 6—then two of them, the total (green peppers) will have to be 6.

Teacher: Ok. What will this (green pepper) be?
Tyreal: Three. Two carrots.

Kim's analysis of this piece of the transcript includes:

> Notice how Tyreal uses algebraic reasoning in his solution. In particular, he establishes the following: 2 carrots = the value 3; 4 carrots = the value 6; 6 carrots = the value 9. His solution represents a one-to-one function. In the course of counting 3. 6. 9 he identifies a pattern and generalizes the relationship of the carrots to the green pepper. He also applies the transitive property of equality in his answer.... He compares carrots to the corn and pepper, and then corn to pepper. He concludes that two carrots equal a green pepper. Moreover, he demonstrates the ability to manipulate symbolic expressions. Tyreal creates a mental representation of 1.50 then, stating one-fifty, Tyreal uses 1.50 mentally and symbolically represents the quantitative values of 1.50 to solve the value of 2 carrots.

The Tug of War problem includes three pictures: the first balances four oxen with five horses. The second drawing balances an elephant with two horses and an ox. The final drawing asks which side of the tug of war would win, an elephant and three horses or four oxen. Although Kim had anticipated that the students would complete this problem readily, as she observed their activity during the lesson, she recognized that they would benefit from using the remaining time in the class period to develop, represent, and share their solutions. In the debriefing discussion following this implementation, all of the observing teachers agreed that this was the appropriate decision.

During the debriefing, Laurie shared her observation of two groups of boys as they struggled to find ways to represent the relationship between the strength of the horse and the oxen and actual exchanges of the animals. She reported that in each case the boys had been stymied by their observation that an ox was worth one and one-fourth horses. Based on real-world experience, they immediately concluded that this was impossible and the horses would be dead. Laurie noted, however, that the boys were able to mathematize the situation by using a ratio of five horses to four oxen and then recognizing they were actually giving value to the relative strengths rather than referring to the actual animals.

On the basis of her later analysis of the video, Kim provides transcript as evidence to document Laurey's earlier observations (ibid, p. 14).

Tyreal: What I did was similar to how Keith—but different—I used different numbers, decimals. Since there are 5 horses and 4 oxen, that means each ox must be equal to at least one horse. That means if you take off the last horse, (from the picture of the ox and horses) it will be equal.
Student: (Viewing Tyreal's solution transparency) Oh—so you split it into fourths?
Tyreal: Yeah—It will be 1.2—or 1 and 1/4—so if you have 1 and 1/4 going to one ox and the horse is 1, the elephant will have to equal up to 3 and 1/4. Then for the last, all you have to do is add them up. You'd have 6 and 1/4 and 5!

Kim, in her analysis of this episode of the videotape, noted:

> Keith and Tyreal's solution work shows the development of mathematical ideas from arithmetic reasoning to algebraic reasoning... Once they established the value of an ox and a horse, they were able to rename the value of the elephant. Keith renamed the strength of the elephant as 13; Tyreal renamed it as 3.25. Eventually both results assisted the students in solving for the group of animals with the greatest strength.

Kim selected another episode from the Tug of War investigation as a critical event because she recognized that it evidenced the students' apparent understanding of equal quantities that could be used interchangeably while also documenting the difficulty that one of the students had in communicating his idea to his partner (ibid, p. 12):

Diesha: We have to figure out because it is 4 ox and 5 horses. But then there is one elephant, one ox and this (two horses). So we can figure out what an elephant is if we figure out the oxen and horses…

Kim: An elephant is as strong as one ox and 2 horses?

Diesha: This is one ox…wait a minute. We need to figure out how much horses weigh… not weigh but like…, You said that 4 ox equals 5 horses, then it's only 3 horses (in the final picture). An elephant is equal to an oxen and 2 horses, so you have 3, 4, 5…that's 5 horses and another oxen…I said there are 4 oxen and 4 oxen equals 5 horses but an elephant is equal to an ox and 2 horses. So you have 5 horses and an ox.

Dr. A: Where did you get the 5 horses and an ox?

Diesha: Because an elephant is equal to 2 horses and an ox.

Juanita: Where did you get 5 horses and an ox?

Diesha: A elephant is equal to 2 horses and an ox.

Juanita: Is that in the picture?

Diesha: It says that an elephant is equal to 2 horses and an ox. So if we take the elephant and replace it with…. Take….

Juanita: Does it say anything about taking away the elephant?

Diesha: But it's just replacing because it is equal to that.

Juanita: How are you replacing the elephant? It says the picture below.

Diesha: It's like replacing because it is the same exact….

In her written summary of her implementation session (ibid, p. 15), Kim documents the value of her subsequent study of the videotape in identifying this and other episodes from the session as important.

> I followed the Powell et al. (2003) model for analyzing video data. I searched for instances of students articulating algebraic ideas to the facilitator, partner or to the entire class. I was interested in students' approaches and how those methods were represented. I looked for significant moments of students' algebraic reasoning. I identified four events that display algebraic (or lack of) reasoning. An event is called critical when it demonstrates a significant or contrasting change from previous understanding, a conceptual leap from earlier understanding.

There were three subsequent implementations in 5th-, 6th-, and 8th-grade classes. These were also videotaped along with the three debriefing discussions. Based on the discussion after Kim's class, the teachers of the final implementations chose to use some combination of the same series of problems. Each teacher stated as a goal that the students should investigate ideas of equality and/or inequality as they constructed solutions for the tasks. Each of the teachers stated as a goal for themselves as facilitators that he or she would carefully attend to how the students were engaging in the activities and determine the amount of time needed for the students to solve and discuss each of the tasks rather than insisting that the students complete the entire series.

Roya, as we noted earlier, had asserted that her students appeared to rely on what she called guess and check as a strategy. Her plan for implementation of the research lesson with one of her 8th-grade classes included a series of tasks beginning with the two Balance Scales problems. As the students became engaged in developing their solutions to the two problems, and Roya realized the constraints of her 40-minute class period, she decided spontaneously to give the entire period to the development and discussion of the solutions to these two problems rather than attempting to rush to cover more material. Her later analysis of the videotape of her implementation evidences her recognition of the variety of student strategies as well as the impact of her actions on the students:

> A majority of the students used some form of direct reasoning. However not all had the same justification and in one case, the correct answer was arrived by chance through faulty reasoning…Aaron and Lekan…solved the problem by assigning weights to the fruits… Danny and Nicole reasoned that since 10 bananas equal 2 pineapples, then 5 bananas equal 1 pineapple…. He then argued that since 5 bananas equal an apple and 2 bananas, if we take 2 bananas from each side of the scale…one apple equals 3 bananas…. Gina and Alejandra set up a wrong proportion which, luckily resulted in the correct answer…. I had the chance to engage in a conversation…to try and understand their reasoning….

After studying the video, Roya included another episode as an important example of one student's, Daiyah's, reasoning, even though it appeared to be very similar to that of Danny and Nicole.

> Teacher: It is more or less the same reasoning (as Danny). Right? What is different? What is different about your way than Danny's way?
> Daiyah: His sounded a lot more confusing than mine.
> Teacher: His sounded a lot more confusing? You tell us what you—what you did.
> Daiyah: I said 10 bananas is equal to 2 pineapples.
> Teacher: So? What was—how did you do it?
> Daiyah: I said 10 bananas is equal to 2 pineapples and divide by—divide both sides by 2. Five bananas equal to one.
> Teacher: Five bananas equal one what?
> Daiyah: Pineapple.
> Teacher: Five bananas equal one pineapple in weight. Okay,
> Daiyah: And then I said that one pineapple equals two bananas and an apple. The apple takes the place of three bananas.

Roya, in her analysis, concludes that:

> Most students in the class had more or less a similar explanation to Danny's but Daiyah states that her justification was a lot less confusing than Danny's. She explains that since five bananas equal an apple plus two bananas, then the apple can be replaced by three bananas. Her reasoning was simpler, more visual and perhaps less sophisticated than Danny's who had used the concept of the Subtraction Property of Equality to eliminate two bananas from each side of the scale.

Conclusions, Implications, and Limitations of the Study

The purpose of this study has been to investigate the four research questions stated earlier. As documented in the results reported here, teachers consistently attended to the mathematical reasoning of the students. Nowhere in the observation notes or analyses of the sessions is there evidence of any of the teachers focusing on classroom management or student behaviors. Rather, the data evidences that throughout the lesson study the teachers were noting and describing specific strategies selected by the students, occasional distracters that students encountered, and differences in the ways that various students approached, represented and solved the problems.

Both Laurey and Kim were able to follow the thinking of the students as they assigned values to vegetables on the balances and the animals in the Tug of War, and to identify these actions as algebraic. A tentative conclusion is that their own engagement in solving the problems themselves and comparing their solution strategies were important preparations for working with the students.

Sarah's course reflection provides some evidence for this conclusion in her statement:

> I personally benefited from working through the problems, getting stuck at times, and having my thinking provoked to try different ways of viewing a problem. Further, as our group shared their solutions to problems, I benefited from hearing and seeing the different approaches used to arrive at the same answer. Actually doing the math, listening to others for understanding, and then discussing the "how's and why's" of different of different approaches helped me to get "unstuck" and try new ways of problem solving.

Roya's early protests about how students appeared to approach problems using a guess and check strategy are in strong contrast to her thoughtful and very specific analysis of the strategies used and mathematical concepts involved in the solutions of her students during the lesson study implementation.

Notice that in each of these cases, the facilitator listened and asked open questions, but did not impose her own ideas on the solutions of the students. In Roya's transcript of Daiyah, the reader can almost feel how difficult it was for Roya not to impose her own opinion that the two solutions, Danny's and Daiyah's, were really the same.

Sarah's self-report about an interaction with two of her students during her implementation adds confirmation to this conclusion about teachers' increasing attention to their students' thinking and to her thoughtfulness about the impact of her interactions as teacher. She notes that two of her students have selected values for the Shorts and Glasses problem that work for the first equation, but not the second:

> When I asked Triana and Carin if those numbers would work if we plugged them into each different combination…they realized that it would not total $50.00 in each. As I reviewed their written explanations later, I saw that the girls had written up a corrected explanation for the price of the items…. This was an example where questioning about the reasonableness of an answer…was helpful.

The specificity and thoughtfulness of the analyses by those teachers who carefully studied the videotapes of the sessions is evidenced by the richness of the data that we have had for this report. Instances such as Kim's description of the solutions to the Tug of War and Roya's understanding of the differences between Danny and Daiyah's solutions are much enhanced, perhaps even made possible, by their opportunity to study students' engagement in mathematics when the teacher was not there. The results provide convincing evidence that teachers' understanding of the complexity of mathematical activity in classrooms where students are encouraged to investigate and construct solutions can be enriched through the collaborative enterprise of lesson study and enhanced by the continued study of the resulting recorded data. The constraint of time that accompanied this particular lesson study cycle, with the final implementations occurring after the course sessions were complete, precluded the opportunity for the group to study any segments of the videotapes together. Evidence from earlier lesson studies in this district and others attests to the value of collaborative video analysis. It is unfortunate that administrative changes in the district and budget constraints eliminated the possibility of continuing the lesson study series that was proposed for the fall when the current groups' analyses of video data could have provided a context for decisions about goals for the next cycle.

It is natural to assume that the depth of understanding for the two teacher-researchers who invested additional hours in structured analysis of the data was greater than that of their colleagues. This differentiation of commitment in no way negates the value for the other participants. Rather, it would seem to provide evidence for the growth of potential leadership for districts. It is gratifying to report that since this experience, each of the coauthors has been promoted to a position of leadership as a mathematics coordinator in her particular district.

One can only conclude from this particular case study that there is measurable value for the participating teachers, their students, and schools in such an intense, thorough, and ongoing approach to professional development. However, even when the implementations are open to colleagues, the numbers of teachers willing and able to commit the time and effort to this endeavor are inevitably small. A number of other district teachers and math coaches attended sessions in which the teachers from the class shared their activities, described their planning process, and discussed what they had learned about the algebra strand in their curriculum. Some of these teachers and coaches attended one or more of the implementations. The value of this limited participation is difficult to determine. Questions about scaling up inevitably lead one to wonder about what will be lost when proposed adaptations include shorter and less frequent preparatory meetings squeezed into the layers of school responsibilities placed on each teacher in addition to their primary task of mathematical instruction for their students.

Evidence from the discussions during the class sessions and debriefings, the examples cited by the teachers in their written reports and analyses, and their reflective comments about the lesson study process document the teachers' increasing ability to recognize and describe the variety of strategies and representations constructed by their students. There is also evidence of the teachers increased awareness of the impact of their actions and interventions on their students' mathematical thinking.

Whether this effort can be sustained over time is undetermined. Districts, especially in urban areas, demand evidence that any expenditure of funds results immediately in increased student performance and test scores. However, the interest on the part of these teachers to continue their analysis and discussion of the results of this experience with the intention of improving their own practice allows one to hope that their commitment may encourage district policy makers to value this observably important activity and to develop opportunities that will provide similar experiences for others.

References

Alston, A. S., Potari, D., & Myrtil, T. (2005). An analysis of teachers' mathematical and pedagogical activity as participants in lesson study. *Proceedings of the Twenty-seventh Annual Conference of the North American Chapter of the International Group for the Psychology of Mathematics Education*. Roanoke, VA.

Alston, A., Maher, C., Pedrick, L., Seffero, M., & Sran, M. (2007). Studying an emerging model of professional development. Poster presentation. *Proceedings of the Twenty-ninth Annual Conference of the North American Chapter of the International Group for the Psychology of Mathematics Education*. Lake Tahoe, NV.

Ball, D. B. & Bass, H. (2003). Making mathematics reasonable in school. In J. Kilpatrick, W. G. Martin, & D. Schifter (Eds.), *A research companion to principles and standards for school mathematics* (pp. 27–44). Reston: NCTM.

Berensen, S. (2004, 2005). External evaluation reports for improving teacher quality through lesson study. Professional development projects of RBDIL, Rutgers University, NJ, USA.

Carpenter, T., Franke, M., & Levi, L. (2003). *Thinking mathematically: Integrating arithmetic and algebra in elementary school*. Portsmouth: Heinemann.

Derry S. (2007). *Guidelines for video research in education: Recommendations from an expert panel*. Chicago: Data Research Development Center (National Science Foundation).

Goldman, R. (2007). *Points of Viewing*. British Columbia: Merlin.

Hill, H., Rowan, B. & Ball, D. (2005). Effects of teachers' mathematical knowledge for teaching on student achievement. *American Educational Research Journal, 42*(2), 371–405.

Laffey, E. H. (2008). *Teacher discourse community: What it says about knowledge for teaching mathematics*. Germany: VDM.

Lewis, C. (2002). *Lesson study: A handbook of teacher-led instructional change*. Philadelphia: Research for Better Schools.

Loucks-Horsley, S., Hewson, P., Love, N., & Stiles, K. (1998). *Designing professional development for teachers of science and mathematics*. Thousand Oaks: Corwin Press.

Maher, C. A. (2005). How students structure their investigations and learn mathematics: Insights from a long-term study. *The Journal of Mathematical Behavior, 24*(1), 1–14.

Maher, C. A. (2008). Video recordings as pedagogical tools in mathematics teacher education. In D. Tirosh and T. Wood (Eds.), *International handbook of mathematics teacher education* (Vol. 2: Tools and processes in mathematics teacher education, pp. 65–83). Rotterdam: Sense Publishers.

Maher, C. A., Muter, E. M., & Kiczek, R. D. (2007). The development of proof making by students. In P. Boero (Ed.), *Theorems and proof in schools: from history, epistemology and cognition to classroom practice* (pp. 197–208). Rotterdam: Sense Publishers (PB ISBN 90-77874-21-6; HB ISBN 90-77874-22-4).

Meyer, M., & Pligge, M. (1998). *Comparing quantities. A unit from mathematics in context*. Chicago: Encyclopedia Britannica.

Morris, K. (2009). *Describing a middle school mathematics lesson study from an eighth grade perspective*. Unpublished master's thesis, Graduate School of Education, Rutgers University, New Brunswick, NJ, USA.

National Council of Teachers of Mathematics (NCTM). (1989). *Curriculum and evaluation standards for school mathematics*. Reston: Author.

National Council of Teachers of Mathematics (NCTM) (2000). *Principles and Standards for School Mathematics*. Reston: Author.

Powell, A. B., Francisco, J. M., & Maher, C. A. (2003). Analytical model for studying the development of learners' mathematical ideas and reasoning using videotape data. *Journal of Mathematical Behavior, 22*(4), 405–435.

Stigler, J., & Hiebert, J. (1999). *The teaching gap.* New York: The Free Press, a division of Simon and Schuster, Inc.

Lesson Study as a Learning Environment for Coaches of Mathematics Teachers

Andrea Knapp, Megan Bomer and Cynthia Moore

Background and Related Literature

Given the emerging knowledge about the benefits of lesson study for developing teachers' knowledge, commitment and community, and learning resources, we chose to employ lesson study as a professional development model for *coaches* of mathematics teachers (see the chapter by Meyer and Wilkerson in this volume for a description of lesson study). The field of professional development coaching currently enjoys steady growth in mathematics education as schools search for effective ways to support the learning of inservice teachers (National Council of Teachers of Mathematics [NCTM] 2000; Stigler and Hiebert 1999). Although coaching is popular as a means of professional development for teachers, its forms of implementation vary from context to context (Olson and Barrett 2004), as do the forms of professional development afforded to coaches. We utilized lesson study to provide coaches and teachers a window into *student thinking* and as a research tool to provide researchers a window into *coaches' and teachers' thinking* with respect to mathematical knowledge for teaching. Throughout this chapter, *teachers* refers to classroom teachers, *coaches* refers to university students hired on fellowship to be content specialist coaches, and *students* refers to K-12 students studying under the teachers.

The context for this study was to utilize lesson study in the professional development of two NSF Graduate STEM Fellows in K-12 Education (GK-12) coaches at Mid-Western University (pseudonym) and one classroom teacher. In the USA, the National Science Foundation GK-12 program awards grants to universities to place college students from disciplines of science, technology, engineering, and mathematics (STEM) into K-12 classrooms as content specialist coaches to jointly design and deliver instruction with classroom teachers. Consequently, the GK-12 program integrates research and teaching through a professional development model (Moore 2003) that is a form of *collaborative coaching* (Olson and Barrett 2004). In this model, both the coaches and the teachers bring knowledge to the table. Coaches

A. Knapp (✉)
The University of Wisconsin-Stevens Point, Stevens Point, WI, USA
e-mail: Andrea.Knapp@uwsp.edu

L. C. Hart et al. (eds.), *Lesson Study Research and Practice in Mathematics Education*,
DOI 10.1007/978-90-481-9941-9_12,

bring content expertise, and they receive training on campus in constructivism and pedagogy consistent with established mathematics and science standards (NSF 2004). Teachers, on the other hand, bring to the partnership knowledge of their students and knowledge of the contextual constraints in which they practice (Cobb and Smith 2007). Teachers voluntarily participate in the GK-12 program to improve their teaching. Thus, this study was inspired by the desire to improve coaches' knowledge about teaching mathematics through lesson study. We asked, *In what ways do participants in a collaborative coaching environment develop mathematical knowledge for teaching as they engage in lesson study?* Specifically, in the environment described here:

1. In what ways do mathematics teachers and coaches develop mathematical knowledge for teaching?
2. In what ways is the development of teachers and coaches related?

Theoretical Framework

We chose to analyze our data using the theory of mathematical knowledge for teaching (MKT) because of its link to student achievement (Hill et al. 2005). Furthermore, the question of how teachers develop essential knowledge for teaching underlies the theory of MKT (Hill et al. 2008). Mathematical knowledge for teaching includes the mathematical knowledge and habits of mind involved in the work of teaching (Hill et al. 2005). More specifically, mathematical knowledge for teaching is categorized in six parts (Hill et al. 2008). Common content knowledge (CCK) is basic, layperson knowledge of mathematical content. Specialized content knowledge (SCK) is the way mathematics arises in classrooms, such as for building representations. An example of SCK is knowing how to represent 1/3 with a diagram (Hill et al. 2005). Knowledge of content and students (KCS) is knowing how students think about mathematics. Knowledge of content and teaching (KCT) involves knowing the most effective examples or teaching sequences. We understand Shulman's (1987) definition of pedagogical content knowledge to be a marriage of KCS with KCT; whereas CCK and SCK represent mathematical knowledge as opposed to pedagogical knowledge (Hill et al. 2005). Knowledge of curriculum and knowledge at the mathematical horizon are the final components of mathematical knowledge for teaching. In this study, we focused on SCK, KCS, KCT, and general pedagogy that fell outside of MKT.

Methods of Data Collection and Analysis

We chose to employ qualitative, multitiered teacher development experiment (TDE) methodology for this study because the purpose for a TDE is to generate models for teachers' mathematical and pedagogical development, closely matching our research aims for the coaches (Lesh and Kelly 2000; Presmeg and Barrett

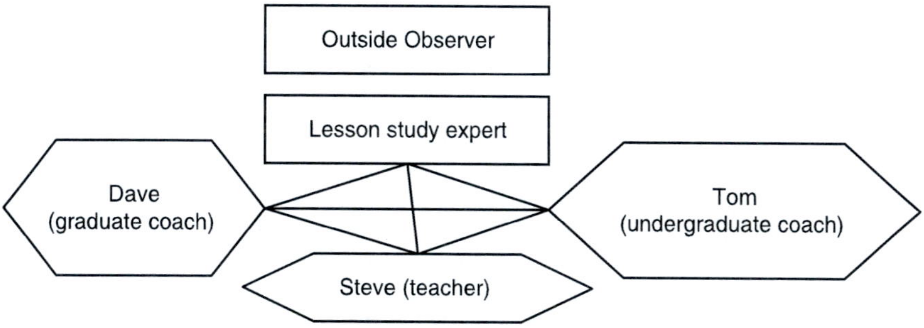

Fig. 1 Lesson study participants

2003). A teacher development experiment is analogous to a teaching experiment with students, but the focus is on the growth and development of a teacher's ability to successfully engage in the work of teaching. Consequently, this study sought to examine the professional development of one graduate coach (Dave), one undergraduate coach (Tom), and one high school algebra teacher (Steve) as they engaged in lesson study. The first author acted as an outside observer while the second author, a participant observer, facilitated the lesson study as the lesson study "expert" (see Fig. 1).

The group met six times for goal-setting and preplanning. Then, the two-day study lesson was taught one time by Steve, the classroom teacher. Finally, Dave, Tom, and Steve met two times for postplanning. Meetings lasted one to two hours. The lesson was not repeated. For data collection, the research lessons were videotaped, and pre- and postplanning sessions were audiotaped and transcribed. Dave, Tom, and Steve were interviewed before and after the lesson study as well. Finally, lesson plans and student work were analyzed.

Qualitative data analysis focused on identifying portions of transcripts providing evidence for the development of and the opportunity to develop the elements of MKT. We read pilot study data collected the prior year to establish a baseline for participants' existing MKT. The first two authors independently coded data for the MKT categories of SCK, KCS, and KCT as well as student understanding, student attitudes, and general pedagogy. We carefully attended to instances of developing MKT versus mere pedagogy. For example, if the teacher learned to shorten the timing or sequence of a learning task in order to keep students from getting restless, we regarded it as pedagogy only. On the other hand, if the timing or sequence issue was related to facilitating discovery about functions, we would code the episode as KCT. To further focus our coding, the categories of SCK, KCS, and KCT were broken into 17 different analysis codes based on the work of teaching (see Hill et al. 2005). Interrater reliability of 83% was reached among coders for identical portions of transcripts. As these elements of MKT were identified, we noted the ways in which particular elements developed (Coffey and Atkinson 1996). We identified a total of 517 episodes of MKT, of which 327 were identified as knowledge of content and teaching (KCT), 151 as knowledge of content and students (KCS), and 39 as specialized content knowledge (SCK). Following the coding process, we compiled the

Table 1 Top ten ways of developing MKT during lesson study process

	Ways MKT developed	Frequency
1	Planning higher-order, open-ended, or real-life questions or tasks; planning questions as coaches practice proposed lessons during preplanning. (KCT, KCS)	56
2	Planning timing and order of activities in the lesson, logistics (KCT, KCS)	52
3	Teacher and coaches sharing common student difficulties, student thinking, representations, lesson ideas, experiences with one another from past experience; teacher teaching coaches what's taught at different grade levels and orders; participant sharing personal learning experience (KCT, KCS)	49
4	Encouraging and figuring out how to incorporate deep, conceptual, student-centered discovery and exploration vs. procedures/terms (KCT)	35
5	Choosing/inventing contextual examples, mathematically better examples or numbers, and alternate representations (KCT)	50
6	Brainstorming lesson ideas and topics to include discussing new content, incorporating mathematically significant ideas, deciding best way to present and use teaching tools; general lesson planning (KCT)	28
7	Introducing, extending, and practicing lessons with technology (KCT, KCS)	25
8	Noting student questions, predictions, products, explanations, responses, and discourse during the lesson study lesson (KCS)	25
9	Considering the mathematics of a representation or example (SCK)	18
10	Encouraging prediction (KCT)	17

Note: Ways with frequency less than 15 omitted

ways MKT developed and looked for themes. We found that some of the ways that KCT developed mirrored the ways that KCS developed. Table 1 summarizes the top ten ways we found MKT to develop. The frequency represents the number of times we identified MKT to develop in the given way. Finally, we coded and compared the pre- and postinterview responses to triangulate our findings.

Results

In this section, we describe the lesson study focus and the ways in which MKT developed as a result of lesson study as a professional development model. Table 1 details the primary ways in which the coaches developed MKT. Table 1 illustrates ways that just one lesson study process, i.e., six planning sessions, two teaching days, and two reflections, facilitated numerous types of MKT development. We then discuss three episodes from the lesson study process that provide evidence for each type of MKT development.

Knowledge of Content and Students (KCS)

The first task of the lesson study group was to identify a practice-based problem or topic for the lesson study. The teacher (Steve) shared student misconceptions which he and other teachers in his department had experienced when teaching slope. During the planning meeting Steve said,

> From the informal survey I took (of a bunch of teachers in our office and from my thoughts) the goals just, students understand that its rise over run. It's move in the y direction first before you move in the x direction which is opposed to being the opposite of an ordered pair going in the x direction first, left and right.

The teacher, Steve, shared with the coaches, Dave and Tom, that students confuse whether slope is a rate of vertical change over horizontal change, or vice versa. After the above discussion, the lesson study group decided to present slope as a unit rate of change through the context of the speed of Thomas the Tank Engine™ toy trains. Thus, as coaches listened to the teaching problems and issues of teachers and other coaches, they developed knowledge of content and students (KCS, Way #3).

Knowledge of Content and Teaching (KCT)

Lesson study provided a safe environment for both teachers and coaches to be vulnerable about their mathematical knowledge for teaching, and thus work collaboratively to support each other's learning. After discussing teaching issues with the teacher, the coaches would often look to research, seek out university mathematics educators, or peruse reform-based curricula for learning tasks which would inform an improved lesson. Coaches would bring these tasks to the lesson study group for consideration, prompting the group to develop knowledge of content and teaching.

As both coaches searched for resources to improve the lessons that they would co-teach with teacher Steve, they became invested in the teaching and learning of mathematics in the teacher's context. Dave, the graduate coach, explained that he and Tom, the undergraduate coach, had made Steve's class "our own". Furthermore, coaches had an opportunity to immediately transfer theory to practice as they implemented, with the input of the teacher, agreed-upon reform-based interventions and supporting technologies they were taught at the university. For example, through discourse during the lesson study process, Dave decided that the goal of teaching slope lacked a contextual basis. Thus, the group added the concept of rate of change to the lesson which they brought to life by using calculator based ranger (CBR) technology (Kwon 2002). The group developed knowledge of content and teaching as they discussed the goals for the lesson (KCT, Way #5). Dave's final interview substantiated this finding.

> Researcher: Okay. So kind of overall big picture, what do you feel like you're taking away from this lesson study?
>
> Dave: I think it made me think about the bigger picture. If I'm teaching Algebra I, what are my goals for the entire course? Break it down, what are my goals for this unit or this chapter? Break it down a little bit more to what are my goals for this lesson or for today? I think sometimes unknowingly the goal is to cover these sections of the textbook.

Dave, who taught two years of high school mathematics before becoming a graduate coach, began to look at his purpose for teaching certain content with fresh eyes. He began to both shift his goals for teaching and also his authority for who decides those goals. He reasoned that he, not the textbook, had the authority to choose goals for his classes.

Teaching Philosophy Tension During Lesson Study

Although Dave saw collaboration with others to plan a lesson as a positive endeavor, he experienced tension through the process as well. Coaches Dave and Tom received instruction about constructivism and the NCTM process standards at Mid-Western University and were expected to implement inquiry-based instruction in the classrooms in which they taught and coached (NCTM 2000). Alternatively, Steve's comfort level lay in teaching traditionally, as evidenced by his initial interview. This disconnect in teaching philosophy manifested as tension during the lesson planning process. In his final interview, Dave said,

Dave:	Also I realized that collaboration is somewhat difficult, especially if you think differently [about teaching paradigms].
Researcher:	Yeah.
Dave:	I think in order to have a lesson study go well, um, you really have to kind of be in agreement on what a good lesson looks like.
Researcher:	Right.
Dave:	Otherwise, it's kind of, I kind of felt like a lot of, not debate, but just talking back and forth and compromise here on this, compromise something else, kind of get it my way when we talk about this. I don't know, it was interesting as far as the dynamics. Not that it was like hard or we didn't get along or anything, we wanted to respect each other and take their ideas, but yet sometimes I felt like I want my ideas implemented too.
Researcher:	Yeah, that's hard. You and Steve I would say definitely would have different types of classroom styles. And so, makes it an interesting group.
Dave:	Right.

Tom, likewise, expressed the tension involved in collaborative planning. He said,

> And so it's interesting; you show a lesson to other people and they go, oh, the one gem of your lesson is not where you thought it was. It's all the way over here. And as you start bringing those different ideas in, you really get to flush out a lesson that's more than just your perspective, and the way that you would just initially rattle off a lesson. It also forces you to think, you know, "Why do I have them putting 10 lines on a page? So I can grade it? You know, why do I want to grade it? Why do I do this?" It has all the good lesson questions you should have asked yourself. The problem is it [collaborating] takes a lot of time. Versus you just sitting out here rattling off what you think a lesson should be and going and presenting it. There's a lot more time, lot more thought process. And also something, that you start losing your part of the lesson. It becomes less and less yours and less and less your voice as more and more input gets put into it.... So your idea ends up wabbling away from you. So the other person is happy, but your little baby is now changed.

Thus, both coaches experienced tension which caused them to critique the teaching styles of the group and to critically examine their own teaching preferences. More important, once the study lesson that incorporated mutually agreed-upon teaching methods transpired, the coaches and teacher were able to scrutinize the success of those methods based on the "court" of the classroom experience (KCT, Way #13). As Dave stated, "[I]t has the potential of kind of forcing Teachers to work together and kind of maybe forcing them to examine what they're doing, like is it really the best?" The immediate testing of strongly held beliefs about teaching appears to be a powerful influence for change inherent in the lesson study process. Although it was uncomfortable at times, the presence of tension indicated that the lesson study process prompted the participants to challenge their implicit beliefs about teaching, thus impacting their knowledge of content and teaching (KCT, Way #4).

Improved Confidence for Preservice Teacher

Before the lesson study, undergraduate coach and preservice teacher Tom, saw his coaching role as that of a "technical advisor" on teaching technologies such as CBRs or *The Geometer's Sketchpad*. In his preinterview, he revealed his lack of confidence in himself as a teacher. He said, "I have a very low confidence value in my teaching expertise, but I have a very high level of confidence inside with the teachers." He tended to not question why something was taught a particular way unless another participant brought it up. He deferred to the teacher as the expert. In the final interview, Tom appeared more confident and excited about teaching. He recognized that the group as a whole was able to improve a lesson. He said, "It's kind of nice to see the fact that not only do we reflect on what the lesson was and how it went, but how can we make it better for next time…." Thus, the collaborative planning and reflection processes appeared to boost Tom's confidence in his abilities to direct instruction. Although improved confidence is not an aspect of knowledge of content and teaching, his improvement in KCT from the collaborative lesson study process clearly impacted his confidence to enact his new knowledge.

New Technology Usage for Classroom Teacher

The lesson study group ultimately settled on teaching the topic of slope to the high school algebra class by running toy trains in front of CBRs. The teacher, Steve, had never used a CBR before. Furthermore, Steve stated that although he had valued technology usage for his advanced classes in the past, he had not considered incorporating it into his entry-level algebra class. Practicing using the technology in the preplanning session prompted the group to anticipate student thinking. In addition, it helped them invent higher-order questions for the lesson which allowed for concrete interaction and student development of concept images for speed and slope (Vinner and Hershkowitz 1980). Thus, not only did the practice time familiarize the teacher with technology uses in the classroom, but it also helped the group invent rich questions for the lesson (KCT, Ways #1, #7) and identify the mathematically significant portions of the activity (KCT, Way #6).

Development of Specialized Content Knowledge

As a result of practicing the lesson with Steve, a newcomer to the technology, the group decided to explicitly address axes before introducing the CBRs. On the first day of the study lesson, the teacher asked the class about independent and dependent variables.

Steve: What was the difference between dependent and independent variable?
Student: Alright, the independent variable, it doesn't have, you don't need it for independent. I can't explain it without an example.

Steve learned from a student response that the students did not have a good understanding of independent and dependent variables (KCS, Way #8). After providing several examples to help the students understand, Steve said to the students, "Label your axis [on whiteboards] with distance and time the way you think it should be listed according to what we've already reviewed. Which axis would be distance, which axis would be time?" Students seemed to be getting the concept that the dependent variable depends on the independent variable. However, at the end of a second day, a student asked a question which threw Steve off. The lesson study group considered the student question in a postplanning session.

Steve: Someone asked, is time always independent? I said 99.9% of the time.
Dave: Always a good answer.
Steve: Leaves a little bit of [room to be wrong].
Dave: For that weird scenario somewhere.
Tom: I'm sure there's some weird physics project where time is a dependent variable.
Steve: That was the kind of thing that popped into my head, so I left open alright maybe there's a tenth of a percent chance of it.

This offhanded student question turned into a discussion involving specialized content knowledge (SCK) during the postplanning session. The discussion later continued.

Dave: If time is your independent variable, it's never going to cross back over itself. Distance versus time. If I'm going from the CBR back, I can get a graph that looks like this. I'm running away, run back, running away, run back. Now if I try to switch it and I say well let's take distance this way and time this way, well at time zero…
Tom: It's not a function.
Dave: It increases. Then at this time I start moving back, and that's the problem. This distance of three feet does not uniquely determine a time, where as time uniquely determines a distance.
Tom: Good answer.
Dave: Where here…I've thought about this a lot. This is a question that I've always had is, "Why can't I switch them?" And it comes down to this. The independent variable…
Tom: Determines the dependent. The dependent is dependent on the independent.

Dave: The independent variable has to vary, independently of anything else.
Steve: Or else it wouldn't be independent. Time can't, it moves forward because you can't move backwards.

The group decided for certain that time could not be dependent on distance. Thus, through reflection afforded by the lesson study process, the coaches and teacher thought deeply about the underlying mathematics in the lesson and began to conceptualize how students come to understand slope and rate of change. They developed specialized content knowledge which is knowledge of mathematics as it arises in the classroom (SCK, Way #9).

Dealing with Cognitive Dissonance

The lesson study trio discussed another scenario that arose during class that showed them that there are some situations in which the independent and dependent axes can be switched. The lesson study group had developed a lesson in which the teacher poured water from one beaker A to another beaker B and asked the students to provide a graph of the situation (KCT, Way #5). Steve was not accustomed to asking such open-ended questions, and he was surprised by the variation in student responses (KCS, Way #8). He had expected students to graph the volume of water in the beaker with respect to time. It caught him off guard when a student graphed the volume of the emptying beaker against the volume of the beaker being filled. The student representation prompted the coaches and teacher to reason that, depending on the viewpoint, either axis could represent either beaker with the amount in beaker A depending on B or the amount in beaker B depending on A. Thus, although the group had studied sophisticated mathematics in their collegiate course work, the lesson study process helped them unpack and more deeply understand relationships between variables in differing contexts, an aspect of specialized content knowledge (SCK, Way #9). This discussion allowed them to make key revisions to the lesson. Moreover, Steve gained experience in running a contextual, student-centered classroom with the support and guidance of the coaches (KCT, Way #4). He said, "It's different for me cause I'm not used to teaching with big props and all that, so it's new for me these last two days."

Thus, the reflection and revision process of lesson study allowed coaches to critique their lesson and questioning with respect to standards- and inquiry-based instruction. Cognitive dissonance, or dissatisfaction with the enaction of the lesson with respect to beliefs about teaching, when experienced within the lesson study environment, allowed coaches to make their implicit beliefs about teaching explicit through discussion and reflection (Olson et al. 2006). The coaches and teacher were then able to purposefully plan to align their teaching practices with their emerging beliefs about the benefits of inquiry-based instruction, thereby improving their knowledge of content and teaching (KCT).

Conclusions

In summary, this study highlighted several ways in which lesson study may support the development of mathematical knowledge for teaching in collaborative coaches and teachers. First, lesson study focused participants on planning open-ended, higher-order questions, especially as they practiced the lesson (KCT). This questioning led to knowledge of content and students (KCS) as students responded in unexpected ways to the content, often with surprising questions for the teacher. The student questions, in turn, prompted the lesson study group to consider mathematical issues related to the context during the revision process, spawning SCK. Thus, all three phases of the lesson study—process-planning, lesson enaction, and reflecting—prompted mathematical knowledge for teaching development. In particular, this study supports the GK-12 expectation that placing STEM content specialist coaches in K-12 classrooms improves the teaching skills of the coaches (Association for the Advancement of Science 2009). For example, the group moved forward in their conception of variable dependency in a time context versus a volume context, affecting the coaches' future teaching abilities to choose tasks appropriately.

The lesson study phases further allowed the researchers to triangulate findings about MKT development as participants engaged in different aspects of the practice of teaching and made connections among those practices. Lesson study as a research tool made participants' thinking transparent as they deliberated about lesson goals, tasks, and revisions. Thus, lesson study served not only as a mediator for participants to learn about teaching and learning, but it served as a mediator for studying coach and teacher knowledge development as well.

Another major finding was that lesson study provided fertile ground for teachers and coaches to explicate their beliefs about teaching and to test those beliefs in a public lesson. The cognitive dissonance that ensued prompted a shift in beliefs away from traditional note-giving to discovery and discourse in the classroom that improved participants' KCT. Finally, lesson study encouraged participants to consider their goals for teaching particular content and to seek out learning trajectories that would foster coherence and mathematical connections.

In response to the second research question, "In what ways is the development of teachers and coaches related?", the development of the coaches and the teacher occurred in tandem. Each participant brought a unique set of experiences and beliefs to the table, and each developed in unique ways. Dave learned that the textbook is not the driving authority for goals in the classroom. Tom gained confidence in lesson development and in his beliefs about inquiry-based teaching. Lastly, Steve shifted toward student-centered teaching.

Implications

We believe that the mathematical knowledge for teaching development of the lesson study participants occurred as a result of the safe, equitable environment provided by lesson study. Lesson study provides the dual benefit of professional development of participants, as well as *nurturing the relationship* of coaches and teachers as equal partners in improving the teaching and learning of mathematics. We feel that coaching and lesson study are not mutually exclusive forms of professional development. The implication for mathematics education is that lesson study provides a learning environment in which coaches can ground their pedagogical and mathematical learning in the context of the inservice teachers that they support. Furthermore, these findings may have implications for preservice teachers' clinical experiences. This finding carries particular importance for teacher education because often preservice teachers abandon the standards-based instruction they receive at the university once they shut the door to their own classrooms. The findings may also have implications for graduate assistants and professors in which content specialists are called on to support teachers or to become teachers themselves. Whereas university course-work may be suitable for development of mathematics content knowledge, lesson study, with its built-in structure of planning and revision, presents itself as a strong model for KCT and KCS development. Finally, we posit lesson study as both a professional development model and as a research tool for studying mathematical knowledge for teaching growth and development. Further study is needed on the impact of lesson study for MKT development of larger numbers of coaches and teachers in varying contexts. This study was supported by NSF grant: DGE-0338188.

References

Association for the Advancement of Science. (2009). *NSF graduate teaching fellows in K-12 education program*. http://nsfgk12.org/. Accessed 19 April 2009.

Cobb, P., & Smith, T. (2007). The challenge of scale: Designing schools and districts as learning organizations for instructional improvement in mathematics. In T. Lambert & L. Wieles (Eds.), *Proceedings of the 29th Annual Meeting of the International Group for the Psychology of Mathematics Education*. Stateline (Lake Tahoe): University of Nevada.

Coffey, A., & Atkinson, P. (1996). *Making sense of qualitative data*. Thousand Oaks: Sage.

Hill, H., Ball, D. L., & Schilling, S. (2008). Unpacking pedagogical content knowledge: Conceptualizing and measuring teachers' topic-specific knowledge of students. *Journal for Research in Mathematics Education, 39*(4), 372–400.

Hill, H. C., Rowan, B., & Ball, D. L. (2005). Effects of teachers' mathematical knowledge for teaching on student achievement. *American Education Research Journal, 42*(2), 371–406.

Kwon, O. (2002). The effect of calculator-based ranger activities on students' graphing ability. *School Science and Mathematics, 102*, 57–67.

Lesh, R., & Kelly, A. (2000). Multitiered teaching experiments. In A. Kelly & R. Lesh (Eds.), *Handbook of research design in mathematics and science education* (pp. 197–230). Mahwah: Lawrence Erlbaum Associates.

Moore, C. J. (2003). *Partnership for research in science and mathematics education mission statement.* Normal: Illinois State University, Biology Department.

National Science Foundation. (2004). *Program solicitation*: NSF 04-533. (2004). The National Science Foundation Web site. http://www.nsf.gov/home/crssprgm/gk12. Accessed 30 September 2004.

NCTM. (2000). *Principles and standards for school mathematics*. Reston: NCTM.

Olson, J., & Barrett, J. (2004). Coaching teachers to implement mathematics reform recommendations. *Mathematics Teacher Education and Development, 6*, 63–78.

Olson, J., Colasanti, M., & Trujillo, K. (2006). Prompting growth for prospective teachers using cognitive dissonance. In J. Novotna, H. Moraova, M. Kratka, & N. Stehlikova (Eds.), *Proceedings of 30th Conference of the International Group for the Psychology of Mathematics Education* (Vol. 4, pp. 281–288). Prague: Charles University.

Presmeg, N., & Barrett, J. (2003). Lesson study characterized as a multi-tiered teaching experiment. In N. Pateman, B. Dougherty, & J. Zilliox (Eds.), *Proceedings of the 27th Conference of the International Group for the Psychology of Mathematics Education* (Vol. 4, pp. 47–54). Honolulu: PME.

Shulman, L. S. (1987). Knowledge and teaching: Foundations of the new reform. *Harvard Education Review, 57*(1), 1–22.

Stigler, J., & Hiebert, J. (1999). *The teaching gap: Best ideas from the world's teachers for improving education in the classroom*. New York: The Free Press/Simon & Schuster.

Vinner, S., & Hershkowitz, R. (1980). Concept images and common cognitive paths in the development of some simple geometrical concepts. In R. Karplus (Ed.), *Proceedings of the Fourth International Conference for the Psychology of Mathematics Education* (pp. 177–184). Berkeley: University of California, Lawrence Hall of Science.

Walking the Talk: Lessons Learned by University Mathematics Methods Instructors Implementing Lesson Study for Their Own Professional Development

Michael Kamen, Debra L. Junk, Stephen Marble, Sandra Cooper, Colleen M. Eddy, Trena L. Wilkerson and Cameron Sawyer

Introduction

As lesson study in the United States has focused primarily on K-12 classrooms, the mathematics education community has embraced lesson study (Fernandez and Moshida 2004; Isoda et al. 2007; King and Murata 2005; Glencoe and McGraw-Hill 2010) in such a way that many mathematics lesson study groups have formed to engage in professional development of both preservice and inservice teachers. The Lesson Study Communities Project in Secondary Mathematics (EDC 2002) and the lesson study groups described by Puchner and Taylor (2006) serve as examples. The later explored a relationship between lesson study and mathematics teachers' self-efficacy. Three universities around the time of this project had adapted lesson study in the university classroom for professional development of university instructors: the University of Wisconsin-La Crosse, Colorado State University, and Southwestern University.

Cerbin and Kopp (n.d.) proposed a model for adapting lesson study to higher education and have implemented this model on their campus. Research on their implementation of lesson study showed that college-level lesson study resulted in teaching improvements. Lesson study at the collegiate level has yielded carefully planned and field-tested lessons for adaptation by other instructors through scholarly inquiry and the formation of communities of practice (Cerbin and Kopp 2006).

Colorado State University has implemented a lesson study approach addressing "professional and curricular development" for calculus instructors. Chappell (2003, p. 1) describes the elements of the project as: "1) collaboratively designing lesson plans that develop the understanding of concepts central to calculus; 2) implementing the lesson plans; and 3) improving plans and instruction based on classroom observations and student achievement on common exams". The report claims that the impact on the quality of teaching and learning is evident and has resulted in the department receiving an award for instructional innovation.

M. Kamen (✉)
Southwestern University, Georgetown, TX, USA
e-mail: kamenm@southwestern.edu

L. C. Hart et al. (eds.), *Lesson Study Research and Practice in Mathematics Education*,
DOI 10.1007/978-90-481-9941-9_13,

At Southwestern University in Georgetown, Texas, Michael Kamen became interested in how to adapt lesson study to the faculty level. A pilot project with colleagues from Southwestern University was implemented that used a collaborative professional development model based on lesson study for structuring peer visitation. The project culminated in the facilitation of lesson study, with four research teams consisting of professors from five academic departments.

Before facilitating lesson study with his colleagues at Southwestern University, Michael set up two lesson study teams with instructors from other universities to experience university-level lesson study as a participant. One of these teams developed a research lesson on teaching preservice teachers about assessment issues during hands-on science instruction, and the other was a research lesson designed to help preservice elementary teachers understand the importance of supporting children's learning with invented mathematics problem-solving strategies. In this chapter the authors focus on the second of these projects, an elementary mathematics methods research lesson.

The Team

Seven professors involved in mathematics education from five universities formed a research team to implement lesson study on their own teaching. Since the individual professors were involved with facilitation of lesson study with preservice and inservice teachers, there was a significant amount of lesson study expertise (as facilitators) among team members. However, none of the team members had participated in a lesson study about *their own teaching*. The initial goal was to experience lesson study as a participant rather than as a facilitator. The team wanted to "walk the talk" by actually trying out the lesson study process that they had been collectively promoting and facilitating.

At the time of the lesson study, Michael, the project initiator, taught K-8 mathematics methods courses at Southwestern University. Cami was a mathematics professor at Southwestern University with a growing interest in mathematics education and joined the team as a content expert. Debbie was an instructor of elementary mathematics methods and graduate curriculum courses at The University of Texas at Austin. Trena was a mathematics education professor at Baylor University, and Colleen was her doctoral student at the time. Sandi was a mathematics education professor at Texas Tech University. Stephen was an education professor and taught in the Masters of Education in Teaching program at the University of Hawaii.

From the beginning, the participants' goal was to experience lesson study first hand. Most of the team had facilitated lesson study with preservice or inservice teachers. The initial expectation was that they would all better understand lesson study from the perspective of a participant and therefore be more effective in facilitating future lesson study activities. The group decided to plan a lesson together and each teach this lesson to their own classes and present both a public lesson and a report on their professional development experience at the 2005 School Science and Mathematics Association (SSMA) national conference.

Several challenges needed to be met in order to implement this plan. Team members came from different colleges and universities in various cities (and even different states in one case). And, they were faced with a deadline for the public lesson just eight weeks from the initial meeting. They met the challenge through a combination of careful scheduling and creative use of web-based technology.

Michael invited colleagues to join the lesson study team and initiated the planning through e-mail which included setting up a schedule for meetings and teaching the lesson, discussing a focus for the lesson study, and sharing the SSMA proposal (for feedback). The team, with the exception of Stephen, attended a face-to-face planning meeting in mid-August at Southwestern University in order to develop the lesson sequence, select tasks, and identify observational areas.

The first two iterations of the lesson were completed in three days, early in September in Austin at The University of Texas. Debbie taught two sections on a Monday, so the first lesson was taught in the morning, the team debriefed at the break, and then the revised lesson was taught in the afternoon. Michael, whose university was just 30 miles north, taught the lesson the following day. This meant that Trena and Colleen from Baylor University and Sandi from Texas Tech University were able to make one trip to participate in the first three iterations. For the next two iterations of the lesson prior to the SSMA conference, Sandi broadcast one of her lessons at Texas Tech using a web-based live video protocol and video recorded another lesson. Because of the travel time and distance, members were not able to physically be present at Texas Tech, thus the use of technology allowed for observation and discussion. The team communicated during the live lesson and debriefed using online chat (instant messaging). The comments made during this on-line discussion as well as the video recording of the second lesson were archived for later review and analysis.

Following the first five iterations, the sixth lesson was taught to a class of local elementary education majors at the 2005 SSMA annual conference. This public lesson study presentation was part of the team's session on lesson study at the conference. At each iteration, a typical lesson study protocol was followed during the debriefing session: the instructor reflected on the lesson first, and then each person in the rest of the group contributed his/her observations.

A critical component of lesson study is that members have the commitment to finding the time and mechanisms to make the process work under whatever circumstances the group faces. The team members were willing to be flexible in their teaching schedules and methods for observing lessons and debriefing lessons. Six iterations of a lesson is certainly not necessary for the lesson study process, but the team pushed the initial goal of "walking the talk," by implementing the process within a variety of settings.

Experiencing Lesson Study: Walking in Our Own Classrooms

To begin the lesson study process, the team chose a common concern: helping preservice teachers (PSTs) better understand the value of allowing children to invent their own strategies. The team also wanted the PSTs to recognize the deep understanding of mathematics needed to appropriately analyze and facilitate dis-

The Critical Role of Task Development in Lesson Study

Brian Doig, Susie Groves and Toshiakira Fujii

Many teachers, researchers and systems have taken the lesson study process (and the related Japanese-style *structured problem-solving* lessons which typically form the basis of the research lesson) to heart, with lesson study groups blossoming around the world (see Hollingsworth and Oliver 2005; Marsigit 2007). While Lewis et al. (2006) claim that lesson study has "come of age" in the United States, underlying assumptions that shape classroom practice can pose obstacles to successfully implementing Japanese-style lessons (and hence also lesson study) in non-Japanese settings (see Doig et al. 2001; Sekiguchi 2005; Groves and Fujii 2008). So, for example, Perry and Lewis (2008) describe a long-term implementation of lesson study in California as "an 'existence proof' that lesson study can be practiced [*sic*], adapted, and sustained by US educators, [that] also highlights the persistent, extended learning by practitioners needed to adapt this form of teacher professional development to the US" (p. 23).

A significant focus for research into lesson study has been *neriage*—the 'kneading' stage of a lesson that allows students to compare, polish and refine solutions through the teacher's probing of student solutions (see Inoue 2008). Frameworks for effective teaching to support children's conceptual understanding also emphasize the need for tasks that are mathematically challenging and significant (Askew et al. 1997). Unfortunately, the report by Hollingsworth et al. (2003), on the *Third International Mathematics and Science Study* (TIMSS) 1999 video study, showed that about three-quarters of the problems set for Australian students were low in procedural complexity and repetitious—in sharp contrast to those problems set for students in higher achieving countries such as Japan.

In earlier research, Australian school principals, teachers and mathematics educators were found to strongly supported classrooms functioning as *communities of inquiry*, while believing that Australian practice falls far short of this goal, partly because the cognitive demands of typical lessons are low and do not challenge children, and partly because of the lack of conceptually focussed, robust tasks that are available to be used to support the development of sophisticated mathematical

B. Doig (✉)
Deakin University, Melbourne, Australia
e-mail: badoig@deakin.edu.au

L. C. Hart et al. (eds.), *Lesson Study Research and Practice in Mathematics Education*,
DOI 10.1007/978-90-481-9941-9_15,

thinking (Doig et al. 2001; Groves and Doig 2002). They argued that a singular feature of a Japanese lesson they were shown that enabled the class to function as a community of inquiry was the nature of the task. This task, which focussed on the concept of a circle, is described later in this chapter.

We argue that Japanese educators place a strong emphasis on task design and selection and that this effort is largely ignored by non-Japanese adapters of lesson study, possibly because the effort involved may be almost invisible, in the way that about 90% of an iceberg is invisible, with all of our attention going to its visible tip.

In this chapter, we focus on *kyozaikenkyu*—the study of instructional materials—and the role of task development in lesson study. We identify and illustrate four types of tasks typically used as the basis for research lessons, discuss the role of lesson study in supporting teachers in their planning of structured problem-solving lessons, and conclude with recommendations for the role of lesson study in teacher education.

Lesson Study and Task Design

Despite records of successful implementation of lesson study as a form of professional development for teachers, we believe that the practice of lesson study outside Japan—although Huang and Li (2008) make a case for China—often tends to overlook the critical role played by the stimulus activity (the task or problem) and its presentation (the *hatsumon*) in creating the foundation of the research lesson, and that attention needs to be paid to "analyzing the topic carefully in accordance with the objective(s) of a lesson" (Shimizu 2002, p. 4). Takahashi (2006) argues that Japanese mathematics lessons emphasize the process of problem-solving activities and provide students with opportunities to re-invent mathematical ideas and concepts by themselves "and this is the reason that lessons employ carefully selected word problems and activities, and their cohesiveness" (p. 3). Further, he points out that "the selection of a problem for the problem-solving activity…is extremely critical for teachers when they plan a lesson" (p. 4). In Perry and Lewis' study (2008), teachers reported, among other changes, "increased use of tasks that elicit student thinking and support student exploration; [and] more experimentation with mathematical tasks before giving them to students, in order to understand task demands and anticipate student thinking" (p. 17).

In Japanese research lessons, the process of selecting the problem or task for the problem-solving activity comes about through *kyozaikenkyu*, which is the investigation of a large range of instructional materials, including textbooks, curriculum materials, lesson plans and reports from other lesson studies, as well as a study of students' prior understandings "which makes it possible for teachers to be able to anticipate students' reactions and solutions to the problems students study during lessons" (Research for Better Schools n.d.). While all teachers need to engage in *kyozaikenkyu* as part of their lesson planning, lesson study requires teachers to engage in it in much more depth.

Watanabe et al. (2008) remind us that the purpose of lesson study is not just to improve a single lesson but to improve mathematical instruction in general, which involves careful attention to *kyozaikenkyu*, something that is not always attended to in non-Japanese lesson study. While the literal meaning of *kyozaikenkyu* is the study or investigation (*kenkyu*) of instructional materials (*kyozai*), the word *kyozai* means much more than textbooks or curriculum materials and needs to involve learning goals. According to Yokosuka (1990)

> It is important that *kyozai* and subject matter content (specific knowledge and procedures to be learned through lessons) are distinguished. It is possible to explore the same subject matter with different *kyozai*, or we can investigate different subject matter with the same *kyozai*. (p. 19, translation cited in Watanabe et al. 2008)

Furthermore, according to Watanabe et al. (2008), "*Kyozaikenkyu*, is the process to help teachers gain a deeper understanding of *kyozai*". It is

> the entire process of research activities related to *kyozai*, beginning with the selection/development, deepening the understanding of the true nature of a particular *kyozai*, planning a lesson with a particular *kyozai* that matches the current state of the students, culminating in the development of an instructional plan. (Yokosuka 1990, p. 73, translation cited in Watanabe et al. 2008)

This notion is also found in the Netherlands in the *Realistic Mathematics Education* (RME) approach (Freudenthal 1973; van den Heuvel-Panhuizen 2001) where a sequence of learning experiences, termed a hypothetical learning trajectory, is posited before tasks are constructed, although this has its difficulties (see Figueiredo et al. 2009).

While there may be insufficient attention to *kyozaikenkyu* in non-Japanese implementations of lesson study, there is limited yet increasing attention being paid to task design in Western contexts. For example, Wiliam (2008) emphasizes the need for mathematical tasks that are both engaging and 'contingent'. In his view, contingency is a key element for sustaining learning, which is not aided by the discrete, unrelated task of many classrooms. Further, Swan, a noted British educational designer, points out that design principles include focussing directly on significant conceptual obstacles and using tasks that are accessible (Swan 2008), while Teppo and van den Heuvel-Panhuizen (2008), presenting reasons for 'mathe-didactical' task designs in mathematics education research, claim that

> not only does an unpacked understanding of the mathematical possibilities (or lack thereof) inherent in the task increase the potential of the research to probe for rich mathematical activity, but this analysis also informs the nature of the inferences that are made related to observed behaviour. (p. 206)

This clearly relates to a lesson as much as to a research endeavour, particularly with respect to the inferences to be made from student behaviour.

Watson and Mason (2007), in a recent review of the ways in which mathematical tasks are used in (Western) mathematics teacher education, suggest that the task

> in the full sense includes the activity which results from learners embarking on a task, including how they alter the task in order to make sense of it, the way in which the teacher directs and redirects learner attention to aspects arising, and how learners are encouraged to reflect or otherwise learn from the experience of engaging in the activity initiated by the task. (p. 207)

This definition of *task* is similar to that of Mok (2004) in which what is "constituted by the interaction between the task and the discourse between the teacher and students or between students…[is] called a 'learning task' event" (p. 2). On the other hand, Herbst (2008) suggests, more succinctly, that the task is "a representation of the *mathematics* to be learned…[and that it is] an opportunity to *study* and *learn* [that] mathematics" (p. 126, italics in the original).

While researchers may disagree over the exact definition of a mathematical task, there has been agreement for some years that the task has a key role in the planning of a lesson (see Brousseau 1997; Doyle 1988). We likewise adopt the view of Japanese educators, where the problem selection, and its presentation to the class, is a distinct characteristic of a lesson (Takahashi 2006). The distinction between task and activity, made by Christiansen and Walther (1986, p. 247), is also worth noting. They also suggest that the "widespread use of ready-made tasks…serves to reduce the teacher's personal investigation to questions about accessibility for *his* students" (p. 249, italics in the original). And this was in the era before Internet lesson-planning sites became available!

Others, such as the influential National Council of Teachers of Mathematics (NCTM) in the United States, have expressed the centrality of the task in effective mathematics classroom practice. The NCTM's position was that "tasks convey messages about what mathematics is and what doing mathematics entails" (National Council of Teachers of Mathematics 1991, p. 24). Similarly, but earlier, Freudenthal (1973) argued for mathematics tasks that adhere to the principle of 'guided re-invention'; that is, tasks that help students to construct new mathematical ideas for themselves.

However, Henningsen and Stein (1997), who cite Doyle's earlier (1988) work on tasks, also point out that the task is necessary but not sufficient. They argue "that the mere presence of high-level mathematical tasks in the classroom will not automatically result in students' engagement in doing mathematics" (p. 527). They state that

> our findings suggest that there was a discernable set of factors influential in assisting students to engage at high levels. These included factors related to the appropriateness of the task for the students and to supportive actions by teachers, such as scaffolding and consistently pressing students to provide meaningful explanations or make meaningful connections. (p. 546)

Nonetheless, we argue that while the centrality of the task *per se* is beyond question, there are other factors that will mediate task effectiveness. For example, we concur with Tsur (2008, p. 141) when he says that the teacher's "facility with using the task as a pedagogical tool" is a factor to which one must pay due attention, thus implying that professional development of teachers is therefore also a vital factor.

Thus, with Doyle (1988), and others, we argue that teachers should pay careful attention to the extent and the way in which, mathematics is emphasized by the task. We characterize the careful attention of "analyzing…carefully" (Shimizu 2002, p. 4) as exploring an iceberg below the waterline, to understand the hidden support that makes it float. That is to say, we need to explicate all the mathematical concepts and understandings that make a particular task or problem 'float' mathematically.

Four Types of Tasks Used in Research Lessons

In this section, we look at four types of tasks typically used in Japanese lesson study research lessons—tasks that

- directly address a concept;
- develop mathematical processes;
- have been chosen based on a rigorous examination of scope and sequence; and
- address known misconceptions.

These tasks have been selected to demonstrate not only different possible pedagogical foci, but also on what the detailed analysis and consideration of the tasks needs to be based—that is: an understanding of the mathematical content; its scope and sequence; children's mathematical understandings and hence their likely responses to the task; children's common misconceptions; and knowledge of a range of tasks and the possibilities the tasks offer to meet the teacher's goals.

The Circle Lesson: Directly Addressing a Concept

This lesson occurred in a Year-3 class of eight children at the Japanese School of Melbourne. According to the teacher, Mr. J, the main mathematics topic for the lesson was 'the concept of a circle'.

The lesson began with Mr. J producing a pole for a game of quoits, where a ring had to be thrown onto the pole from a distance. Mr. J placed the pole in the centre of an open space in the classroom and asked three children (shown as B1, G1 and B2 in Fig. 1) to stand at the three marked places on a red line at one side of the room. Children were concerned that the game would be unfair, but they focused mainly on the distance between children standing along the red line, rather than their distance from the pole.

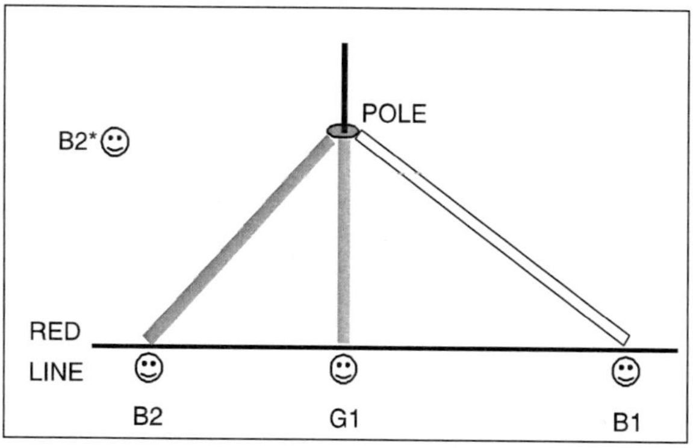

Fig. 1 B2's solution for making the quoits game fair

After a discussion on how to measure distances, with children using a metre ruler to measure their distance from the pole, Mr. J put different coloured, pre-cut strips of paper on the floor, as shown in Fig. 1, he then held them up together to show that their lengths, and hence the children's distances from the pole, were different. He then stated the question for the lesson (the *hatsumon*) as: "How can we make the game fair?" It took five more minutes, during which time children continued to try to find points to stand on the red line, before B2 came up with a way that two people could be the same distance from the pole—he moved the end of the G1's yellow strip to the point B2* on Fig. 1.

Mr. J then gave all the children a yellow strip and asked them to "think for themselves" and find somewhere to stand so that everyone was the same distance from the pole. Children were excited that no-one would be at a disadvantage. This segment took 20 minutes of the 45 minute lesson.

Mr. J then reproduced the situation on a large sheet of paper on the blackboard. He stuck a miniature pole on the paper and asked children to use sticky yellow paper strips and dots to represent their positions on the floor (see Fig. 2).

Mr. J: Look at the different positions—what do you notice?
G2: It's like a round circle [makes circle shape with hands].
G3: No—it's like a flower.
G1: If you follow the end of each yellow strip it will become a circle [traces large circle on the desk with her finger].
Mr. J: What if every student in the school took part? [adds more strips]...
B3: If there are many students standing round, maybe it's a circle.

Mr. J removed the pole and put another sheet of paper over the first with a circle drawn where the dots were and asked "How many yellow points would we need? A hundred? A million?" He then put the word circle on the paper and elicited names for the centre, radius and diameter from the children. The remaining 15 minutes of the lesson were taken up with the children working in pairs drawing circles. Initially, many children chose to use a compass, even though Mr. J told them that they

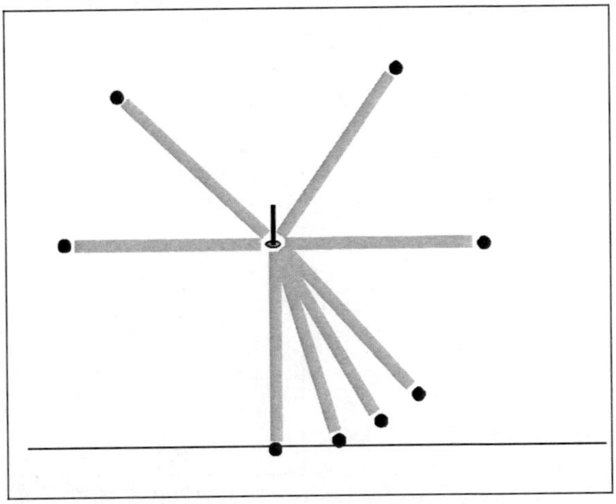

Fig. 2 Paper representations of children's positions

had not yet learned how to use one, and encouraged one girl who said that she could use a yellow strip of paper or a plastic circle to trace round, to show him how. After about 7 minutes, Mr. J asked children to find a way to draw a circle without a compass. A few minutes later Mr. J said: "Now everyone is tracing—is there another way?" Children tried various ways while Mr. J pivoted one of the yellow strips of paper around an end held by his finger. B2 excitedly cried out that he could do it and demonstrated drawing a circle by holding the middle of one end of his pencil case and tracing a circle with his finger in the hole at the other end. Children applauded and Mr. J demonstrated B2's method at the front of the class. The lesson finished with a few minutes of suggestions from children how to fix one end, culminating in the use of a drawing pin. Mr. J summed up by saying: "As you suggested, there are other ways of drawing a circle than just using a compass."

Both in the lesson and in his responses to a questionnaire, it was clear that Mr. J chose the task to enable him to directly address the concept of a circle and that this was consistent with his goals for the lesson. He stated his aims as "children have the *concept* of a circle and find *real* circular objects' [emphasis in original transcript]". According to Mr. J, the most important aspect of the lesson in terms of children's learning was that they understood that the circle is a locus. The purpose of working in groups (in this case pairs) was "to facilitate discussions while working", while the purpose of the whole class discussion was for children to "share ideas and strategies for solutions [demonstrating that] there are many different ways of thinking which reach the same conclusions". Mr. J further described his mathematics lessons in general as follows:

> Introductory lessons [to a topic] use materials. So this was typical. The introduction is very important and takes a lot of time. After that there is much practice, then we go to calculations—a series of 3 or 4 lessons [per topic].

Mr. J concluded with the comment that "Mathematics should be part of children's daily lives." In the 40 minute quoit activity, Mr. J embedded the concept of a circle in a rich, intriguing, intrinsically motivating, problematic framework by asking: "How can we make the game fair?"

This lesson highlights two important features of some Japanese lessons and the tasks that underpin them: the highly conceptual nature of the goal for the lesson and the need for sophisticated mathematical understanding. By way of contrast, in Australia, the concept of a circle as the locus of a set of points would not be regarded as part of the scope of primary school mathematics. Instead, there is considerable emphasis on descriptions and categorization of shapes.

Snow White and the Seven Dwarfs: Developing Mathematical Processes

This lesson took place in a preparatory class in Melbourne, Australia in the middle of the five- and six-year-old children's first year at school. Children sat on the floor while the teacher, Mrs. B, reminded the class that they had heard the story of *Snow White and the Seven Dwarfs* the previous day. She then put out a sheet of paper to represent a 'long table' at which Snow White and the seven dwarfs sat for their dinner. She said that Snow White always sat at the head of the table, while the dwarfs sat at the two long sides, with a different number of dwarfs sitting on each side each day. Seven counters were used to represent the dwarfs. One child was asked to illustrate a possible way—she placed one counter on one side and the remaining six on the other side. The teacher then presented the problem for the day on the board as shown below.

How might the 7 dwarfs sit at the table?

Did you find all the ways?

How do you know?

Children were told they could paste coloured rectangles onto paper to represent the table and draw "quick maths drawings"[1]—not ones with the dwarfs' "hair and hats and eyelashes"—as well as write numbers. Alternatively, they could use concrete materials and jotters, where they could record their answers because "your job is to find as many ways as possible".

The children worked individually or in pairs at their tables or on the floor for about 15 minutes, after which the teacher called the children back to the floor for a discussion of the different solutions.

Mrs. B commented that one child had said he had found seven ways, while another had found six. She reminded them that they wanted to find all the possible ways. As individual children contributed different ways, she wrote their solutions on a piece of card which she attached to a whiteboard as shown in Fig. 3.

The teacher commented that it was very difficult to see whether all the ways were represented, and suggested that they look at an ordering that had been used by one of the girls, Melody (see Fig. 4). When asked to try to find a pattern, the children replied that "it's just the opposite".

After some discussion with Melody as to whether she actually moved the counters around (which she had done) or just wrote the numbers 'the other way around', the teacher noted that there was "another pattern we could make". When none of the children volunteered such a pattern, she showed the class the second pattern that Melody had made—see Fig. 5.

[1] In this description, quotes are from the video recording of the lesson. All children's names are pseudonyms.

Fig. 3 Children's different seating arrangements for the seven dwarfs

2 and 5	1 and 6
7 and 0	0 and 7
4 and 3	5 and 2
3 and 4	
6 and 1	

Mrs. B then reproduced Melody's ordering on the board by writing the first two arrangements on the board and asking different children to supply the remainder: "She's making a pattern—0 and 7, 1 and 6.... Have a look what's happening—0, 1—what do you think comes next?" When the children got to 3 and 4, the teacher asked Ivy, "What is the pattern on this [right] side?" and Ivy replied, "The pattern is 7, 6, 5, 4—like counting backwards...from 7." After asking another child for the pattern on the left hand side, the teacher and children completed the list of arrangements. Mrs. B asked the children whether there was anything different they could have done and, just to make sure, she told the children to go back to their tables and tell her if they had an arrangement that was missing from the list—but to make sure there were still seven dwarfs! Some children thought they had different arrangements, but of course none really did.

Mrs. B then told the children to come back to the floor and asked, "Have we found all the ways?" to which the children chorused, "Yes!" She continued: "Nobody else got any more at their tables for seven...but how do we *know* we've got all the ways?" One boy replied: "We've used all the numbers." The teacher confirmed this, discussed with the children why there could not be more, and asked them again how many ways there were for seven dwarfs. She then asked what would happen if

4 and 3	0 and 7
3 and 4	7 and 0
5 and 2	1 and 6
2 and 5	6 and 1

Fig. 4 Melody's first ordering of the seating arrangements

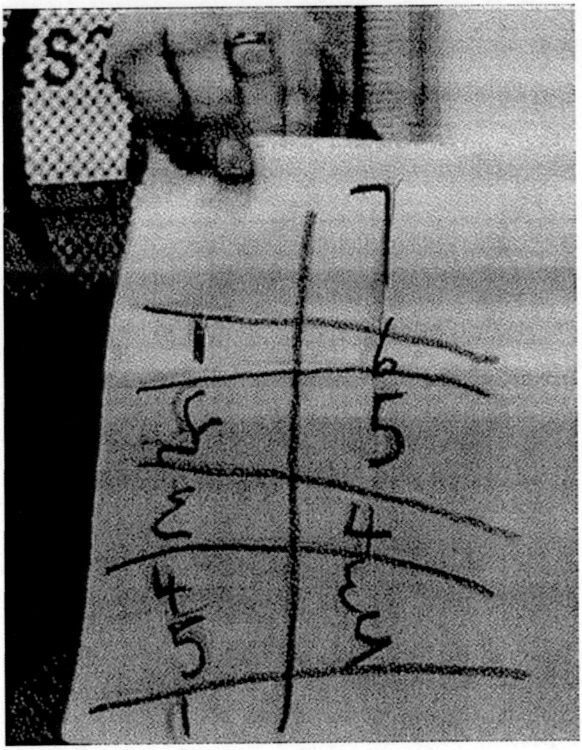

Fig. 5 Melody's second ordering of the seating arrangements

"at the three pig's house" there were eight people—how many different ways could they sit? After Ivy answered 9, the teacher asked what if there were 10 visitors and Caitlin replied 11 ways. "So what if we had all 24 children in the class sitting at a very long table?" Ivy answered 25. The discussion continued:

Mrs. B:	What is the pattern? How did you know each time without doing it? When there were 7 people there were 8 ways. When there were 8 people there were 9 ways. When there were 10 people there were 11 ways. When there were 24 there were... 25 ways. What's the pattern?... How did you know without doing it each time?... What if there 100 people?
Child:	101.
Mrs. B:	What if there were 300 people?
Children:	400, 500, 104.
Caitlin:	301.

Children came to the board to try to write 301—the fourth child wrote it correctly after the first three wrote 1ε1, 3001, and 131, and the lesson ended with a discussion about which of these was correct. In total, discussion of the children's solutions, how they knew they had found all the ways, and the discussion of the number of ways for different numbers of people, took approximately 25 minutes.

When asked about the goals for this lesson, Mrs. B focussed on developing problem-solving processes, such as working systematically, as well as looking for patterns. In Japan, such a lesson might be termed a 'jump-in lesson' to indicate that it could take place at many different points in the curriculum sequence.

The 13 – 9 Subtraction Lesson: A Rigorous Examination of Scope and Sequence

The 13 – 9 subtraction lesson described in this section was part of a sequence of lessons on subtraction with regrouping in a Year-1 class in Japan conducted by a 'veteran teacher' Mr. N. The lesson began with the problem, "There are 13 persimmons. I have eaten □ of them, how many are left?" The teacher started with the number eaten as 2, then 3, then moved on to 9, at which point some children responded that they could not subtract 9 from 3. The teacher stated that the problem for the day was 13 – 9. Children worked individually on the problem for about 10 minutes, during which time the teacher identified three different solution strategies used by the children to be discussed in the *neriage* phase of the lesson. The first strategy was counting down. The second was *subtraction-subtraction* (see Fig. 7) and the third was *subtraction-addition* (see Fig. 6). During this part of the lesson, individual children came up to the blackboard at the front of the class to

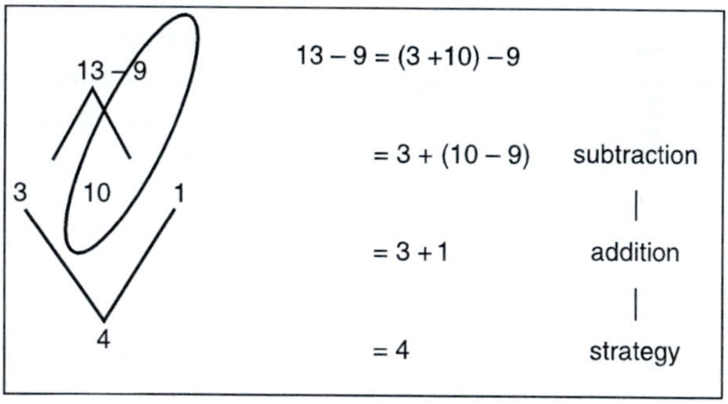

Fig. 6 The subtraction-addition strategy

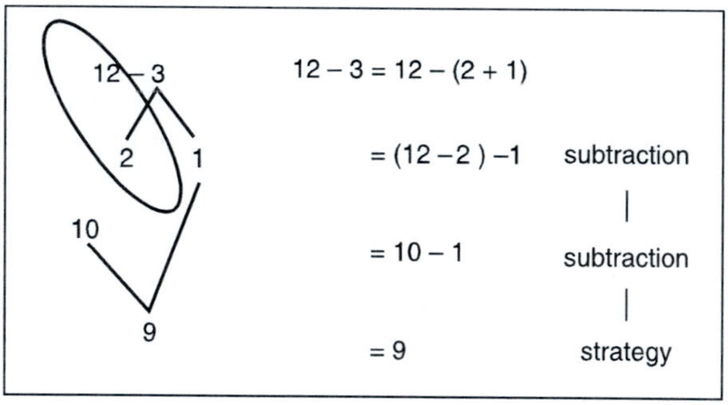

Fig. 7 The subtraction-subtraction strategy

explain their own, or another child's, strategy or to demonstrate it with magnetic blocks.

At the end of the explanation of the subtraction-addition method, the teacher asked each child to use their own blocks to demonstrate this method step-by-step. Although all three methods were explained, the lesson emphasized the subtraction-addition method by letting children experience this method with concrete materials. During the last part of the *neriage* phase the teacher asked children the similarity between the last two methods. Children responded that both methods used 10 as a unity. Children also said that they could use their previous knowledge.

Japanese first-grade textbooks contain a unit concerning subtraction of one-digit numbers from two-digit numbers, with regrouping, which is regarded as an important area of content to learn. There are a total of 36 such possible examples: 18 – 9, 17 – 9, 17 – 8, 16 – 9, 16 – 8, 16 – 7,..., 11 – 3 and 11 – 2. In Japan, which of the 36 tasks should be the first for children to learn is hotly contested. There are six companies publishing textbooks for elementary mathematics in Japan, and all textbooks start the unit with either 13 – 9 or 12 – 9. The second task is 14 – 8, and the third one is 12 – 3. Non-Japanese readers might be surprised that every textbook uses almost the same sequence of tasks, and also that, according to the teachers' guide, one hour should be spent on each of these three tasks.

Why should 13 – 9 or 12 – 9 be the first task for children to learn when they meet subtraction with regrouping for the first time? The reason given is that these tasks naturally lead to the subtraction-addition strategy. The subtraction-addition strategy, illustrated in Fig. 6, refers to subtracting 9 from 10 first, then adding 1 to 3, so the sequence of operations is subtracting first, then adding.

To subtract 9 from 10, we need to see 13 as 10 and 3 from our knowledge of the base ten notation. As we use the base ten system, children have learned to compose, or decompose, numbers with a 10, and in the first-grade textbooks this occurs just before the unit on subtraction with regrouping. The subtraction-addition strategy is usually dominant when solving a problem such as 13–9, because 9 is close enough to 10 for children to naturally relate it to 10, and see 9 and 1 become 10. Therefore the dominant strategy for 13 – 9 becomes $(10 + 3) - 9 = (10 - 9) + 3$ – that is the subtraction-addition strategy.

On the other hand, for the problem 12 – 3, the dominant strategy used by children is the *subtraction-subtraction* strategy. Here, as shown in Fig. 7, the 3 becomes 2 and 1, then $12 - 3 = 12 - (2 + 1) = 12 - 2 - 1 = 9$. This strategy does not depend on place value notation. In this sense, it does not make full use of the base ten system. And also, the subtraction-subtraction strategy has the disadvantage that children often make mistakes in the tens or hundreds places if they try to adopt this strategy for larger numbers.

Consistent with the sequence of tasks for subtraction, the addition part of the textbook uses 9 + 4 as the first task.

When children solve 14 – 8, the two strategies, subtraction-addition and subtraction-subtraction tend to both occur in approximately equal numbers. Therefore, teachers who wish to promote argumentation in their classes sometimes use 14 – 8 as the first task for children, while textbook companies adopt a more conservative stance based on their desire to make it easy for teachers to anticipate student responses and to be sure that there will be enough children who use the subtraction-addition strategy.

Area and Perimeter: Addressing Misconceptions

Space does not permit a full discussion here of the fourth category of tasks, addressing a common misconception. An example of such a misconception, that exists among both Japanese and Western students, is that shapes with the same perimeter have the same area, and so, for example, if we were trying to measure the approximate area of a puddle we could carefully place a string around its perimeter and then deform its shape to that of a rectangle to calculate its area.

The first lesson on area in a fourth-grade Japanese textbook (Hironaka et al. 2006, pp. 22–23) shows children trying to decide which of the two shapes representing two 'newsletters' being compared has the larger area. Shapes have been chosen deliberately to have the same perimeter but different areas. After introducing the idea that area can be measured using square centimetres, a 'maths story' is used to further illustrate the fact that not all shapes with equal perimeters have equal area (see also Takahashi 2006). This addresses the common misconception of there being a unique relationship between area and perimeter for all shapes.

The Role of Lesson Study

While none of the Japanese lessons described here arose directly from lesson study, they are nonetheless the products of lesson study. That is to say, like many lessons in Japanese schools, these lessons owe their focus on mathematics, and their pedagogy of implementation, to teachers who have been imbued with the philosophy and practice of lesson study. This level of professional support, offered to Japanese teachers through lesson study, contrasts dramatically with that available to Australian and American teachers.

Firstly, most Japanese teachers have experienced lesson study themselves. Moreover, they do not have to develop structured problem-solving lessons by themselves—instead they are encouraged, like Mr. J did, to adapt lessons from the textbook, or draw on published articles such as the 'Study on teaching materials' or 'Practical study' sections of the *Journal of Japan Society of Mathematical Education*. In the case of research lessons for lesson study (and at other times too, presumably) teachers can usually also draw on the expertise of an expert mathematics teacher in their school. All of these support teachers in their *kyozaikenkyu*.

In Australia there have been a number of projects which have resulted in the publication in hard copy or online of so-called exemplary lessons (see Lovitt and Clarke 1985). However, for all but the most competent teachers, there remain problems with using such "exemplary" materials. Many of the lessons are designed to be highly flexible and capable of being adapted for a wide range of year levels. Often links with the 'regular' curriculum are not obvious and they are more of the Japanese 'jump-in' type. When asked about the source of her tasks, Mrs. B (the Australian teacher in the Snow White example) said she used a wide variety of sources for tasks, including books and the Internet, and frequently adapted tasks or developed her own. She stated that "There's no one particular place I get things from, I wish there was because it would make it a lot easier."

As with other researchers (see Swan 2008; Takahasi 2006, for example), Mrs. B's main criteria for suitable tasks were that all the children

- can start [the task] and that the solution is not immediately obvious;
- are interested, that they are engaged with the story, or whatever [the task was];
- have choice with whatever they use to help them solve the problem.

For Mrs. B, the selection of task is influenced by contextual factors as well as the mathematics. She is a well-known, highly experienced and highly respected teacher, with an excellent understanding of the scope and sequence of mathematics, particularly in the first few years of school, but without the support of a tradition of lesson study she is very much left to work on her own in carrying out *kyozaikenkyu.*

Regarding the subtraction lesson, Western observers are often astonished not only by the thought of entire lessons being spent on single tasks like these, but also by the order of presentation being the subject of so much study and debate. However, Japanese lesson study is frequently used to investigate sequences of tasks that are different from those traditionally used. For example, while one of the authors was in Japan recently, she observed several of a sequence of eight or more research lessons designed to explore the effect of introducing quotitive (measurement) division before partition (sharing) division to a Year-3 class. Thus, Japanese lesson study involves continuing efforts to examine and improve approaches: it is not seeking merely to transmit a single 'best' approach to all teachers.

The area and perimeter lesson, mentioned only briefly here, illustrates the use of extensive research into finding and using tasks that directly address common student misconceptions. Teachers need to be aware of the research literature on children's understandings, and use this knowledge when planning lessons and selecting tasks.

The 'iceberg' metaphor reminds us that there are many aspects of mathematics tasks to be considered when planning a lesson. For example, as described above, selecting a suitable task, or activity, in the lesson study context is not simply a matter of finding a task that carries the required mathematical content. The fact that different versions of the task may provoke different responses from the children necessitates a careful choice to be made. This is evident in the lesson introducing subtraction 'across ten', 13 – 9. Clearly it is imperative for a teacher to know the common strategies that children use, in order to orchestrate the discussion of a range of strategies. According to Killen (2009), being prepared for a variety of responses is a key point in any lesson involving discussion.

Further, selecting the exact wording of the *hatsumon*, or question posed to the children to solve, is also a critical step in the planning of a lesson. To many, the question of how to make the quoit game fair could seem a long way from the lesson intention of exploring the concept of a circle. However, as we have described, this outcome was achieved in a very engaging and, we would hope, memorable manner. Again, in the Snow White example, the question posed was intriguing and appropriate for the age group. Moreover, it was an opportunity to allow children to explore and use a mathematical process, as well as to experience, in school, the unusual situation of there being a variety of correct answers.

This is not to argue that other aspects of the Japanese lesson are unimportant: the discussion, polishing and refining of children's strategies (*neriage*), observing differences in children's work (*kikan-shido*), and summarizing (*matome*) are also indispensable supports to an effective lesson (Shimizu 1999). However, we wish to concentrate, in this chapter, on the more 'hidden' facets of lesson study, ones that support the iceberg: these we wish to place in the forefront of our thinking, in order to gain the greatest benefit for those involved in the practice of lesson study in countries outside Japan.

The development of practice that already has some local currency is thought to be the most effective way in which to change teacher practice (following the I'Ching argument that going with the river's current rather than against it, is more effective). While the precise manner in which these practices would be developed, and culturally mediated, is still currently a work-in-progress, our research on *Communities of Mathematical Inquiry* (see Doig et al. 2001; Groves et al. 2000; Groves and Doig 2004), suggests that many teachers are aware of deficiencies in current practice, and are eager for professional development that addresses these. The situation in the United States is more advanced than Australia, with more than 500 active lesson study groups. Thus, it is suggested that following the example of the United States, of mathematics educators working closely with small groups of interested teachers, offers opportunities for creating a cadre of effective teachers with clear understandings of not only what to do, but also how to achieve effective mathematics classroom learning: teachers who are attentive to the tasks that they use, and capable of "analyzing…carefully" (Shimizu 2002, p. 4) these tasks.

Agreeing with Zaslavsky (2007, p. 435) that "effective MTEs [mathematics teacher educators] engage MTs [mathematics teachers] in carefully crafted tasks in order to…construct what they need to know about teaching school mathematics", we believe that research into "carefully crafted" tasks suitable for teacher education is a priority no less than is research on lesson study tasks for classroom teachers. There are at least two key reasons for this priority. First, this research would provide the information needed to change and improve the learning of mathematics pedagogy. Second, such research would provide tasks for mathematics teacher educators that would allow them to model the type of classroom pedagogy that we believe our mathematics teachers should be using. Although examples of suitable tasks have existed for some time (see Lovitt and Clarke 1985), their use has suffered to date, in Tsur's (2008) terms, from teachers' inability to use tasks effectively as a pedagogical tool.

Clearly, we need to heed Hiebert and Stigler's (2000) argument that "improving teaching does not depend on eventually perfecting 182 lessons but rather on engaging intensively with the issues involved in teaching any lesson" (p. 16). Such intensive engagement, in both teacher pre- and in-service education, we believe, should provide future teachers with practical experience of effective mathematics pedagogy, as well as a better understanding of the foundations of this pedagogy and the mathematics that they are preparing to teach: lesson study has the potential to do this.

However, if the key elements of lesson study are to be effective more widely as Professional Development for teachers, a re-conceptualization of curriculum and textbooks is needed, to assist in re-orientating teachers, and researchers, to the need for coherent sequences of lessons which are focussed on the mathematical tasks *per se*. While examples of curriculum materials with such an orientation exist, they are not widely seen outside the Netherlands (see van Galen et al. 2008) or Japan.

Conclusion

We firmly believe that the practice of lesson study, either as part of initial teacher education, or later professional development, has the potential to increase the number of effective mathematical learning experiences enjoyed by children. But lesson study is not achieved without effort: it is no 'silver bullet'. The reason for this is that the strengths of lesson study rest on two significant bases. The first is the detailed planning of lessons, which, in turn, is based on deep reflection on the mathematics and the pedagogy. While many lesson study groups outside Japan focus attention on the mathematics, often this is at the expense of the pedagogy, or *vice versa*. It is critical that a balance be maintained.

The second basis of lesson study is cultural, including both the classroom culture and the wider professional culture of teachers. In an ideal lesson study classroom, the 'didactic contract' assures that every student willingly engages in the set tasks, contributes to discussion, and knows that their contributions are valued. This is not always the case in non-Japanese classroom cultures. Additionally, in the broader professional culture of teachers, lesson study requires a culture where being open to other perspectives on teaching and critical commentaries on a lesson are seen as positive contributions to pedagogical knowledge and understanding. For many non-Japanese teachers this is not an easy attitude to achieve.

However, despite such difficulties, cultural and pedagogical, we believe that lesson study, in its full sense, has the potential to make learning mathematics an enjoyable and positive experience for students, as well as a rewarding professional experience for teachers. Further, we hope that the examples and discussion provided in this chapter will help to reveal possibilities and encourage teachers to consider exploring beneath the iceberg.

Acknowledgements *Mathematics classrooms functioning as communities of inquiry: Models of primary practice* was funded by the Australian Research Council. The Chief Investigators were Susie Groves (Deakin University), Brian Doig and Laurance Splitter (The Australian Council for Educational Research).

Talking across cultures was funded by the Deakin University *Quality Learning Research Priority Area*. The project team was Susie Groves, Brian Doig (Deakin University), Toshiakira Fujii (Tokyo Gakugei University), Yoshinori Shimizu (Tsukuba University), and Julianna Szendrei (Eötvös Loránd University, Budapest).

References

Askew, M., Brown, M., Rhodes, V., Johnson, D., & Wiliam, D. (1997). *Effective teachers of numeracy. Final report.* London: King's College.

Brousseau, G. (1997). *Theory of didactical situations in mathematics* (Didactique des mathematiques, 1970–1990 N. Balacheff, M. Cooper, R. Sutherland, & V. Warfield, Trans.). Dordrecht: Kluwer Academic Publishers.

Christiansen, B., & Walther, G. (1986). Task and activity. In B. Christiansen, G. Howson, & M. Otte (Eds.), *Perspectives on mathematics education* (pp. 243–307). Dordrecht: Reidel.

Doig, B., Groves, S., & Splitter, L. (2001). Primary mathematics practice. In J. Bobis, B. Perry, & M. Mitchelmore (Eds.), *Proceedings of the 24th Annual Conference of the Mathematics Education Research Group of Australasia* (Vol. I, pp. 178–185). Sydney: MERGA.

Doyle, W. (1988). Work in mathematics classes: The context of students' thinking during instruction. *Educational Psychologist, 23*, 167–180.

Figueiredo, N., van Galen, F., & Gravemeijer, K. (2009). The actor's and the observer's point of view: A geometry applet as an example. *Educational Designer, 1*(3). http://www.educationaldesigner.org/ed/volume1/issue3/article10. Accessed 13 November 2010.

Freudenthal, H. (1973). *Mathematics as an educational task.* Dordrecht: Reidel.

Groves, S., & Doig, B. (2002). Developing conceptual understanding: the role of the task in communities of mathematical inquiry. In A. D. Cockburn & E. Nardi (Eds.), *Proceedings of the 26th conference of the international group for the psychology of mathematics education* (Vol. III, pp. 25–32). Norwich, UK: University of East Anglia.

Groves, S., & Doig, B. (2004). Progressive discourse in mathematics classes: The task of the teacher. In M. Haines & A. Fuglestad (Eds.), *Proceedings of the 28th Conference of the International Group for the Psychology of Mathematics Education* (Vol. 2, pp. 483–502). Bergen: PME.

Groves, S., & Fujii, T. (2008). Progressive discourse, *neriage*, and some underlying assumptions. *Proceedings of the 41st Annual Conference of the Japanese Society for Mathematics Education* (pp. 525–530). Tsukuba: JSME.

Groves, S., Doig, B., & Splitter, L. (2000). Mathematics classrooms functioning as communities of inquiry: Possibilities and constraints for changing practice. In T. Nakahara & M. Koyama (Eds.), *Proceedings of the 24th Conference of the International Group for the Psychology of Mathematics Education* (Vol. III, pp. 1–8). Hiroshima: PME.

Henningsen, M., & Stein, M. K. (1997). Mathematical tasks and student cognition: Classroom-based factors that support and inhibit high-level mathematical thinking and reasoning. *Journal for Research in Mathematics Education, 28*(5), 524–549.

Herbst, P. (2008). *The teacher and the task.* In O. Figueras, J. L. Cortina, S. Alatorre, T. Rojano, & A. Sepulveda (Eds.), *Proceedings of the 32nd Conference of the International Group for the Psychology of Mathematics Education* (Vol. I, pp. 125–131). Morelia: PME.

Hiebert, J., & Stigler, J. W. (2000). A proposal for improving classroom teaching: Lessons from the TIMSS video study. *The Elementary School Journal, 101*(1), 1–20.

Hironaka, H., Sugiyama, Y. and 36 professors and teachers (Eds.), (2006). *Mathematics 4B for Elementary Schools.* (M. Yoshida, A. Takahasi, T. Watanabe, W. C. Jackson III, M. N. Leer, & J. L. Shouffler, Trans.). Tokyo: Tokyo Shoseki.

Hollingsworth, H., & Oliver, D. (2005). Lesson study: A professional learning model that actually makes a difference. In J. Mousley, L. Bragg, & C. Campbell (Eds.), *Mathematics—Celebrating Achievement. Proceedings of 2005 MAV conference* (pp. 168–175). Melbourne: MAV. http://www.mav.vic.edu.au/pd/confs/2005/. Accessed May 2009.

Hollingsworth, H., Lokan, J., & McCrae, B. (2003). *Teaching mathematics in Australia: Results from the TIMSS 1999 video study*. Melbourne: The Australian Council for Educational Research.

Huang, R., & Li, Y. (2008). *Developing exemplary lessons to pursue mathematics classroom instruction excellence in China.* In O. Figueras, J. L. Cortina, S. Alatorre, T. Rojano, & A. Sepulveda (Eds.), *Proceedings of the 32nd Conference of the International Group for the Psychology of Mathematics Education* (Vol. I, pp. 179–183). Morelia: PME.

Inoue, N. (2008). Zen and the art of 'neriage': Facilitating consensus building in mathematical inquiry lessons. In O. Figueras, J. L. Cortina, S. Alatorre, T. Rojano, & A. Sepulveda (Eds.), *Proceedings of the 32nd Conference of the International Group for the Psychology of Mathematics Education* (Vol. I, p. 275). Morelia: PME.

Killen, R. (2009). *Effective teaching strategies.* Sydney: Pearson.

Lewis, C., Perry, R., Hurd, J., & O'Connel, l., M. (2006). Lesson study comes of age in North America. *Phi Delta Kappan, 88*(4), 273–281.

Lovitt, C., & Clarke, D. (1985). *Mathematics Curriculum and Teaching Programme (MCTP) Activity Banks* (Vol. I and II). Melbourne: Curriculum Corporation.

Marsigit. (2007). Mathematics teachers' professional development through lesson study in Indonesia. *Eurasia Journal of Mathematics, Science & Technology Education, 3*(2), 141–144.

Mok, I. A. C. (2004, April). *Learning tasks.* Paper presented at the Lesson events as the basis for international comparisons of classroom practice symposium at the Annual Meeting of the American Educational Research Association, San Diego. http://extranet.edfac.unimelb.edu.au/DSME/lps/assets/Ida_Learning_Task_AERA2004.pdf. Accessed May 2009.

National Council of Teachers of Mathematics. (1991). *Professional Standards for Teaching Mathematics.* Reston: Author.

Perry, R., & Lewis, C. (2008, March). What is successful adaptation of lesson study in the U.S.? *Journal of Educational Change, 10*(4), 365–391. http://www.springerlink.com/content/vk-20104381w61501/. Accessed May 2009.

Research for Better Schools. (n.d.). *Glossary of lesson study terms.* http://www.rbs.org/lcsson_study/glossary.php#study. Accessed January 2009.

Sekiguchi, Y. (2005). Development of mathematical norms in an eighth-grade Japanese classroom. In H. L. Chick & J. L. Vincent (Eds.), *Proceedings of the 29th Conference of the International Group for the Psychology of Mathematics Education* (Vol. 4, pp. 153–160). Melbourne: PME.

Shimizu, Y. (1999). Aspects of mathematical teacher education in Japan: Focusing on the teachers' roles. *Journal of Mathematics teacher Education, 2*, 107–116.

Shimizu, Y. (2002, May). *Capturing the structure of Japanese mathematics lessons: Some findings of the international comparative studies.* Paper presented at the ICMI-Second East Asia Regional Conference on Mathematics Education and Ninth Southeast Asian Conference on Mathematics Education, Singapore: National Institute of Education. http://extranet.edfac.unimelb.edu.au/DSME/lps/assets/EARCOME%202002%20-%20Shimizu.pdf. Accessed May 2009

Swan, M. (2008). A designer speaks [Electronic Version]. *Educational Designer, 1.* http://www.educationaldesigner.org/ed/. Accessed 23 Sept 2008.

Takahashi, A. (2006). Characteristics of Japanese mathematics lessons. http://www.criced.tsukuba.ac.jp/math/sympo_2006/takahashi.pdf. Accessed Sept 2008.

Teppo, A., & van den Heuvel-Panhuizen, M. (2008). *Qualitative research methods: mathe-didactical analysis of task design*. In O. Figueras, J. L. Cortina, S. Alatorre, T. Rojano, & A. Sepulveda (Eds.), *Proceedings of the 32nd Conference of the International Group for the Psychology of Mathematics Education* (Vol. I, pp. 205–208). Morelia: PME.

Tsur, R. (2008). *A researcher perplexity: Why do mathematical tasks undergo metamorphosis in teachers hands?* In O. Figueras, J. L. Cortina, S. Alatorre, T. Rojano, & A. Sepulveda (Eds.), *Proceedings of the 32nd Conference of the International Group for the Psychology of Mathematics Education* (Vol. I, pp. 139–146). Morelia: PME.

van den Heuvel-Panhuizen, M. (2001). Realistic mathematics education in the Netherlands. In J. Anghileri (Ed.), *Principles and practices in arithmetic teaching* (pp. 49–63). Buckingham, UK: Open University Press.

van Galen, F., Feijs, E., Figueiredo, N., Gravemeijer, K., van Herpen, E., & Keijzer, R. (2008). *Fractions, percentages, decimals and proportions: A leaning-teaching trajectory for grades 4, 5 and 6*. Rotterdam: Sense Publishers.

Watanabe, T., Takahashi, A., & Yoshida, M. (2008). Kyozaikenkyu: A critical step for conducting effective lesson study and beyond. In F. Arbaugh & P. M. Taylor (Eds.), *Inquiry into Mathematics Teacher Education*. Association of Mathematics Teacher Educators (AMTE) Monograph Series, Vol. 5

Watson, A., & Mason, J. (2007). Taken-as-shared: A review of common assumptions about mathematical tasks in teacher education. *Journal of Mathematics Teacher Education, 10*, 205–215.

Wiliam, D. (2008). When is assessment learning oriented? *EARLI/Nofrthumbria Assessment Conference*. Nofrthumbria.

Yokosuka, K. (1990). *Jugyokenkyu yougo daijiten. (Dictionary of lesson study terms)* Tokyo, Japan: Tokyo Shoseki.

Zaslavsky, O. (2007). Mathematics-related tasks, teacher education, and teacher educators. *Journal of Mathematics Teacher Education, 10*(4), 433–440.

The Intersection of Lesson Study and Design Research: A 3-D Visualization Development Project for the Elementary Mathematics Curriculum

Jacqueline Sack and Irma Vazquez

Introduction

This chapter describes how the research methods for developing a 3-D visualization program for elementary children closely resemble the principles of lesson study. While many lesson study experiences offer teachers opportunities for personal professional development to deepen their pedagogical content knowledge, this team's immediate focus is on students' mathematical understanding as they engage in research lesson activities that are the basis for development of new materials for the elementary mathematics classroom.

Mathematics educators sometimes struggle with ways to present research findings to classroom practitioners meaningfully. To this end, in this chapter, the existence of strong methodological connections between research and practice is demonstrated. Marilyn Cochran-Smith (2003) uses the term *inquiry stance* as a construct for life-long teacher development around student learning appropriate to local contexts. The lesson study protocol is a method for developing and sustaining an inquiry stance for teachers. The design research protocol is this research team's method for sustaining an inquiry stance as researchers and as teachers.

Based on the lesson study cycle presented by Murata (see the chapter by Murata, this volume), the two protocols share characteristics in that the three research team members (1) consider goals for student learning and development; (2) plan research lessons based on these goals; (3) during each research lesson, collect data on student learning and development; (4) use these data to reflect on the lesson and on instruction more broadly; and, (5) if desired, revise and reteach the research lesson to a new group of students. The two protocols differ in that generally lesson study intends to focus on teacher development through the lens of student understanding and learning through the use of various instructional approaches, while this team's design research process intends to focus on the impact of newly designed activities on student understanding and learning using established social-constructivist instructional practices.

J. Sack (✉)
University of Houston Downtown, Houston, TX, USA
e-mail: sackj@uhd.edu

L. C. Hart et al. (eds.), *Lesson Study Research and Practice in Mathematics Education*,
DOI 10.1007/978-90-481-9941-9_16, © Springer Science+Business Media B.V. 2011

In the following sections, the theoretical framework that undergirds the development of instructional goals, the classroom context and research methodology, and examples culled from research data that illuminate the close relationship between this team's research method and lesson study will be presented.

Theoretical Frameworks

Spatial Visualization

The National Research Council's (NRC) report, *Learning to Think Spatially* (2006), identifies spatial thinking as a significant gap in the K-12 curriculum, which NRC claims is presumed throughout the curriculum but is formally and systematically taught nowhere. NRC believes that spatial thinking *is* the start of successful thinking and problem solving, an integral part of mathematical and scientific literacy. The National Council of Teachers of Mathematics' (2000) *Principles and Standards for School Mathematics* supports this view. In their early years of schooling, students should develop visualization skills through hands-on experiences with a variety of geometric objects and use technology to dynamically transform simulations of two- and three-dimensional objects. Later, they should analyse and draw perspective views, count component parts, and describe attributes that cannot be seen but can be inferred. Students need to learn to physically and mentally transform objects in systematic ways as they develop spatial knowledge. From a purely academic perspective, researchers who have examined students' performance in higher-level mathematics emphasize the importance of visual processing. For example, Tall et al. (2001) found that to be successful in abstract axiomatic mathematics, students should be proficient in both symbolic and visual cognition; Dreyfus (1991) calls for integration across algebraic, visual, and verbal abilities; and Presmeg (1992) believes that imagistic processing is an essential component in one's development of abstraction and generalization.

Spatial Operational Capacity Framework

The spatial operation capacity (SOC) framework that guides the study is based on the research work of Yakimanskaya (1991), van Niekerk (1997), and Sack and van Niekerk (2009), and aligns with the standards established by NCTM (2000). Children are exposed to activities that require them to act on a variety of physical and mental objects and transformations to develop the skills necessary for solving spatial problems (Yakimanskaya 1991). The SOC framework (see Fig. 1) uses (1) full-scale models (or scaled-down models) of large objects which can be handled by the child; (2) conventional graphic models that are two-dimensional graphic representations which resemble the real three-dimensional objects; and (3) semiotic

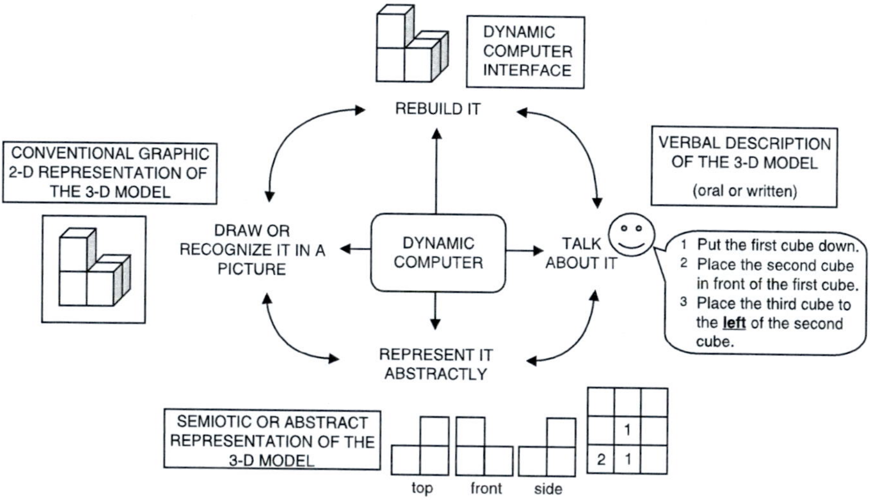

Fig. 1 Spatial operational capacity (SOC) framework

models, such as view and numeric top-view diagrams, which are abstract, symbolic representations (Freudenthal 1991) which usually do not bear any resemblance to the actual objects. Regardless of the representation given in any particular problem, activities are designed to foster children's spatial competence using all of these visual representation modes in addition to verbal descriptions, using physical and mental processes. Transformations are implicitly and explicitly utilized as children engage in activities across the SOC framework. The instructional activities have children moving to and fro (indicated by the double-ended arrows) among these different representations initially using loose wooden cubes and Soma pieces (see Fig. 2) progressing to 2-D conventional graphic and semiotic models and verbal representations integrated with Geocadabra (Lecluse 2005), a dynamic computer interface that was not available when the SOC framework was originally developed. Through the Geocadabra Construction Box modules, complex multi-cube structures can be viewed as two-dimensional representations or in symbolic ways, for example, as top, side and front views or numeric top-view diagrams. Whereas one can move around a three-dimensional model to see it from other vantage points, one may see various views of a computer-generated figure through its ability to be rotated in real time. The computer interface serves as a mediator of knowledge (Borba and Villarreal 2005) rather than as a unique form of representation in that

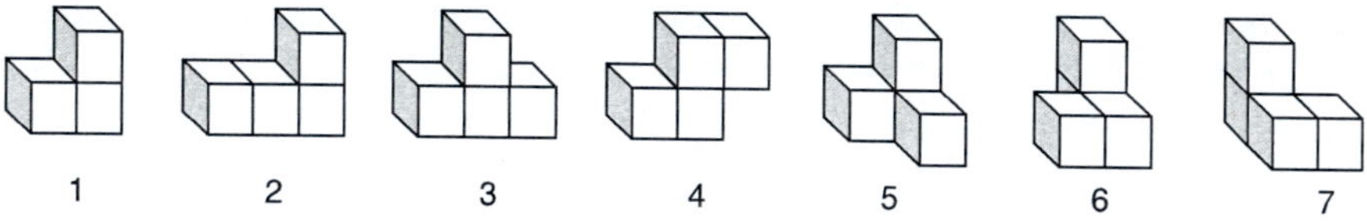

Fig. 2 The complete set of Soma figures

the figures on the computer screen may be viewed statically as in conventional 2-D pictures, or dynamically as one views 3-D real life models.

Context and Methodology

The ongoing study is being conducted in a dual-language urban elementary school serving approximately four hundred students within one of the largest public school districts in the mid-southwestern United States. Approximately 70% of the students are Hispanic, 20% are African American and the remaining 10% are White or Asian. Three-quarters of the students are designated "At-Risk" and 55% are English-language learners. The participants in the study represent a typical cross-section of the larger school community. Mathematics instruction is conducted in Spanish for students below 5th grade during the academic day but this project is conducted in English within an established after-school program at the school.

The ongoing research/instructional team consists of university-based Jackie, who has over fifteen years of classroom experience, and Raquel and Irma, who teach full-time in the school's dual-language program and collaborate closely during the school day on matters related to their academic programs. Irma and Jackie have teamed as instructors for multi-session teacher professional development courses several times. The team shares social constructivist instructional approaches, which are the norm in Irma's and Raquel's classrooms. Students work independently or in small groups of two to four and all students are expected to ask each other for help or support before asking the teacher. Mutual respect is fostered in the classroom environment in which students are comfortable expressing their understandings knowing they are safe to express their confusion or frustration in front of their peers. They are expected to explain and provide justification for their mathematical conclusions. The teacher serves as the lesson facilitator and guide while the students construct meaning and representations for themselves.

For the first year of the study, at the beginning of the fall semester, English and Spanish parent/guardian and student consent-to-participate forms were sent home to parents of all after-school 3rd and 4th graders. All respondents are accepted into the program. Fourteen 4th graders and eleven 3rd graders returned signed consent forms. However, various conflicting activities resulted in attrition and approximately eleven 4th graders and eight 3rd graders attended the program consistently throughout the year. Irma had taught mathematics and science to all 4th-grade participants during their entire 3rd-grade year. During this particular year, staffing changes for the 3rd-grade class occurred and she taught all core subjects to half of the school's 3rd-grade students. Consequently, some of the 3rd-grade participants in the after-school research program were not her students during the school day. However, within the after-school research program, all participants became attuned to her behavioral and communal expectations very quickly.

Each lesson is part of a design research experiment (Cobb et al. 2003) in which the research team hypothesizes learning outcomes, designs instructional activities and materials, based on the SOC framework to support the outcomes, and enacts the lessons with 3rd-grade and then 4th-grade children. During most lessons, the children are encouraged to work with a partner or in a small group. They may choose to work independently unless interaction is expected. For Geocadabra-based activities, they work independently using laptop computers and may interact with those working close by.

The research team also attends to differentiated instructional "processes and procedures that ensure effective learning for varied individuals" (Tomlinson and McTighe 2006, p. 3). The design of each learning experiment is intended to challenge each child according to his or her particular readiness, interest and learning profile. This is achieved by providing open-ended problem situations having multiple solution pathways and many solution options. Most activities involve tasks in which the children reproduce or create figures using at least one of the SOC representations. For example, some activities require them to select from a wide variety of task cards with 2-D pictures comprising Soma figure or loose cube combinations that vary in complexity. Or, they may create figures using Soma figures or loose cubes or using the computer interface and draw the corresponding numeric top-view representations.

One member of the research team takes the instructional lead during each research lesson while the others support and interject if needed. The teacher rarely gives away answers but presses students to resolve their questions and deepen their mathematical understanding through the teacher's scaffolding questions. Each lesson is video recorded by one member of the research team. Video-recorded data consist of whole-class interactions, individual children working on specific activities, and informal interviews between teacher and child or children to make sense of their mathematical understanding.

Each week, after the two classes, the research team meets to discuss their interpretations to ensure that their notes reflect joint interpretations of children's understandings and misunderstandings. In addition, the team uses copies of student work and printouts of their computer work and meeting notes to triangulate among themes that emerge during analysis of these data. During these retrospective analyses, the research team determines the actual outcome of the lesson and then plans the next lesson, which may be an iteration of that lesson to improve the outcome, a rejection of the lesson for future use if it failed to produce adequate progress toward the desired outcome, or a change in direction if unexpected, but interesting outcomes arose that were deemed worthy of further exploration. Lesson plan development emerges out of team discussion about how newly created activities or materials could logically support and develop student understanding and conceptualization resulting from prior lessons. This aspect forms an important and time-consuming process in conventional lesson study in which participant teachers negotiate over a variety of lesson resources that are brought to the table. However, in this research study, hypothetical learning trajectories develop from week to week based on children's interest in an aspect of spatial learning or on how the team views the logical flow of learning about a spatial concept. The research team easily reaches consensus decisions since activities are under development by the team rather than

Table 1 Summary of exemplified SOC stimuli and products

Example	Stimulus	Intermediary product/stimulus	Desired end product
1	Verbal		3-D model (loose cubes)
	3-D model (loose cubes or Soma figures)		Verbal
	Verbal		3-D model (loose cubes or Soma figures)
2	2-D picture	3-D (loose cubes)	Top-view conventional coding (computer)
	2-D picture		Top-view conventional coding (computer)
3	2-D picture (two-Soma assembly)	3-D assembly model (Soma figures)	Top-view conventional or invented coding
	2-D picture (two-Soma assembly)		Top-view conventional or invented coding
	Top-view conventional or invented coding	3-D model (loose cubes)	3-D assembly model (Soma figures)
	Top-view conventional or invented coding		3-D assembly model (Soma figures)

available from a variety of sources. The university-based researcher is responsible for creating and revising new materials, such as task cards, as needed.

In the next section, three example lessons follow for the purpose of demonstrating how the team's research process aligns with lesson study. Table 1 provides a summary of the example activities presented in this chapter along with the SOC stimuli and resulting student-generated SOC products including intermediary 3-D products that served as bridging stimuli to produce the desired end products.

Example Research Lessons

Example 1: Children's Use of Descriptive Language

Preprogram interviews (September 11, 2007) were conducted by two of the teacher-researchers with the 11 3rd-grade and 14 4th-grade students who returned consent-to-participate forms to assess students' initial abilities to perform relatively simple tasks within the SOC framework using loose cubes or single Soma pieces. Items developed for the interviews were based on activities developed by van Niekerk (1997). These included matching 2-D and 3-D Soma figures in identical and isometrically transformed orientations, reproduction of Soma figures using loose cubes, reproduction of Soma figures from verbal directions, and students' drawings of 3-D Soma figures presented in specific orientations. During these interviews, the research team noted some of the unconventional ways that children responded to the tasks with verbal directions to produce 3-D structures using unmarked loose cubes (see Fig. 3) (Sack and Vazquez 2008). Using the example item,

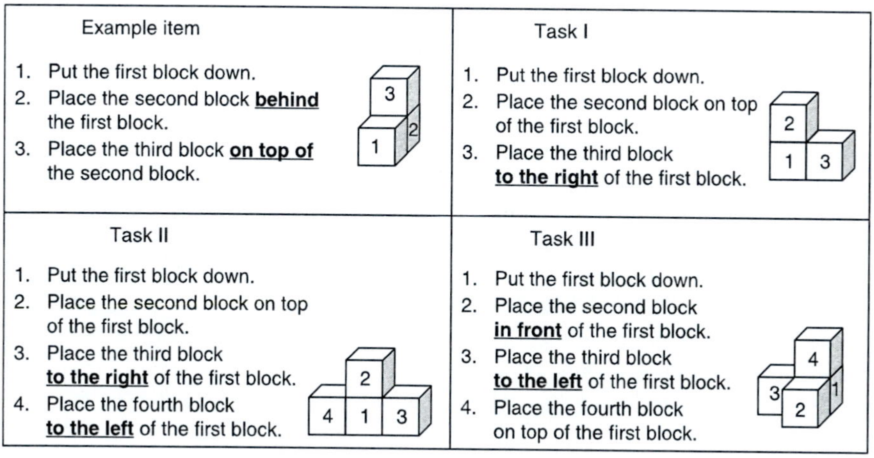

Fig. 3 Preinterview verbal stimuli

the teacher-researcher asked students to build a figure using unmarked, loose cubes while the teacher-researcher read the verbal directions to them and also while the interviewer read with them. Each child selected to listen or to read along for tasks I, II and III.

If a child was unable to build tasks I or II successfully, then she or he was not given the opportunity to build subsequent figures in this section of the preinterview. Only one child from each grade level was unable to reach task III. Of those who attempted task III, 7 out of 10 3rd graders and 3 out of 13 4th-grade students consistently placed the second block behind the first. Of note, when building the figure for the example item which stated, "Place the second block behind the first block," children who erred on task III were also consistent in placing the second block in front of the first. (Sack and Vazquez 2008). The common mathematical convention is to name the face closest to the observer the front face, and that which is furthest away the back face. The research team theorized that the child who sees himself in the referent position will refer to the face corresponding to his own front as the front (as in a translation of position).

During *Lesson 1* (September 18, 2007), the children were exposed to orientation activities in which they built as many 4-cube figures as they could imagine based on a contextual story about designing modular homes out of 4 cubic rooms for an outer-space village. The objective was to build all combinations of 4-cube structures (joined face-to-face) without holes or overhangs and be able to distinguish between those that were images through vertical reflections.

For *Lesson 2* (September 25, 2007), the research team intended to establish the appropriate mathematical conventions for the terms, *in front* and *behind* or *at the back* that moved between the SOC framework's 3-D and verbal representations based on the preinterview result that indicated relatively widespread nonconventional usage of these descriptive terms. Although the team intended the children to discover and develop mathematical ideas on their own, this represents an instance where teaching of mathematical convention takes precedence over invented definition. The plan was to engage the whole class in discussion using appropriate examples and then facilitate activities that required the children to use this language appropriately. During the lesson, three or four children were invited to line up one-behind-another facing the board. The rest of the class sat on the floor also facing the board. The seated audience were asked, "Who is in front?" and "Who is behind?" Then, the line of children turned to face the seated audience and the question was repeated. The child in front facing the board became the child at the back when the line turned. Then, four identical large wooden blocks were lined up and the following questions were asked: "Which is the front block?" and "Which is behind?" The research team hypothesized that the children would appropriately define "in front" and "behind" for in-line figures that have no designated front.

Two activities followed that whole-class discussion. The facilitator modeled the process for each activity by directing two children through the instructional sequence while the others observed. During the first activity, children sat side-by-side in pairs with a divider to hide each other's work. Child A created a 3-D figure with four cubes (like the ones they had created during Lesson 1) and described the structure, step-by-step including the language "in front" and "behind." Child B built the figure using loose cubes from Child A's verbal directions. After removing the divider, the children compared their structures. Then they switched roles. The goal of using appropriate conventional language was not realized during the first hour's lesson. Between classes, the research team restructured the lesson for Grade 4, who attended during the second hour. Debriefing field notes, the research team's record of research lesson reflections, describe the situation:

> Grade 3 had [difficulty] with this [first] activity [so] it was restructured for Grade 4. Children built their figures and described them as they built them based on the language with which they were comfortable. For example, they used left and right arrangements if they did not feel [confident] about differentiating between front and back. Also, if a shape [had reflectional] symmetry, like Soma #3, then even if the partner made errors, she or he could [still] construct the same shape. Children were using the language, but sometimes incorrectly. We decided to eliminate this [activity] for 4th grade and for future [use].

In the second activity, still working with a partner and using a divider, Child A chose a Soma piece and described it orally while Child B built it with loose cubes. They continued as in the first activity. If needed, Child A also built the figure with loose cubes while looking at the Soma model. This activity was successful and those who needed more challenge worked with combinations of two Soma pieces. In this way, differentiated learning experiences were provided according to readiness and interest. Some children continued to confuse forward and backward positional language, often choosing to avoid these terms. Several months later when discussing this lesson with other researchers, the research team reflected that a better intervention might be to ask the children to think about which is the front or back of a house from the perspective of a person who stands on the sidewalk in front of the house.

In summary, this lesson was planned the week before its enactment paying particular attention to how the teacher would enable the children to develop and use conventional positional language. The problems with the first activity of Lesson 2 were recognized during the 3rd-grade session through careful attention to student behavior. Adjustments to the original plan were made for the 4th-grade session that followed immediately that day. The research team agreed, as noted in debriefing notes that both groups performed well on the second activity in which students gave each other appropriate verbal directions for reproducing Soma figures of their own choosing. The research team decided to reuse and reevaluate this activity with a new cohort of 3rd graders the following year. The lesson study components, (1) consider goals for student learning and development; (2) plan a research lesson based on these goals; (3) during the research lesson, collect data on student learning and development; (4) use these data to reflect on the lesson and on instruction more broadly; and (5) if desired, revise and reteach the research lesson to a new group of students, were clearly evident in this research lesson.

Example 2: Revisiting Children's Progress in Coding from a 2-D Conventional Figure

This example illustrates how the research team modified a learning trajectory to meet individual children's needs. The research team designed and created sets of task cards showing 2-D pictures of combinations of two Soma figures. Some sets were printed on color-coded cardstock with one Soma figure shaded to enable easy discrimination between the two Soma figures. For example, all purple task cards had combinations of Soma figure #2 in combination with any of the seven (including #2) as the other Soma figure. Another set, printed on bright pink card stock, consisted of all combinations with one Soma figure shaded. The most complex set was printed on blue cardstock with neither Soma figure shaded. For an example, see Fig. 4.

Through October 2007, these task cards were used in activities in which the overarching objective was to identify the two Soma figures (2-D representation) on each task card and reproduce it using two Soma figures (3-D representation). Most children in both classes performed these tasks with ease rapidly moving to the most comprehensive and complex set of non-color-coded task cards in which neither Soma figure was shaded, with easy and difficult combinations mixed together. Multiple solutions were possible with many of the figures in the most complex set of cards. The children self-selected shaded or nonshaded task cards based on self-perceived levels of interest and ability. The instructor challenged the children to find more than one solution, and to justify why no more solutions were possible if

Fig. 4 Two-figure Soma assembly task cards with and without shading

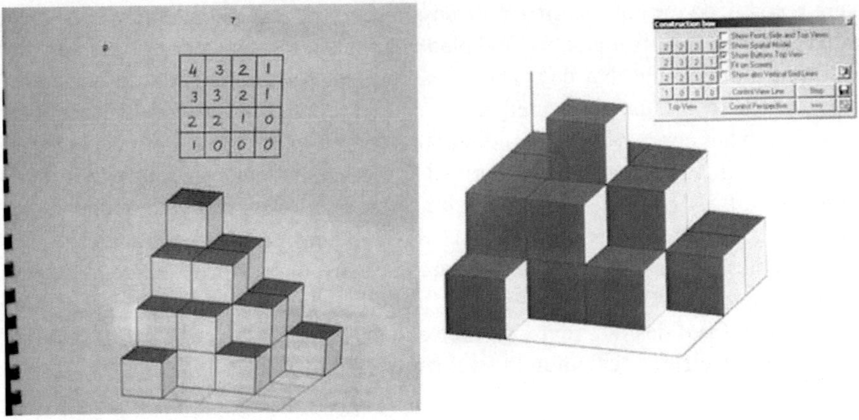

Fig. 5 The figure for Task 1f and Vena's early attempt to reproduce it on the Geocadabra Construction Box

they moved through the activity too quickly. Strugglers soon achieved mastery by first building the figures on the task cards using loose cubes.

The children were introduced to the Geocadabra Construction Box on October 30, 2007. They initially rebuilt the 2-D figures shown in a customized manual (van Niekerk 2006) using loose cubes (3-D representation) and then created the complete structure on the computer using Geocadabra's Construction Box module by clicking on the four-by-four number grid. For an example, see Fig. 5. Ideally, one should correlate the number of stacked cubes on a space with the cardinal number in the corresponding grid space (numeric top-view coding representation). For example, the back row on the "3-D" figure corresponds with the top row on the grid.

The research team noted that some children, Vena in particular, had failed to make the mental connection between a stack's spatial position (2-D figure) and its corresponding grid position (abstract coding), incorporating a transformation (rotation) that implicitly required mental imaging. The following excerpt from field notes, Lesson 8 (November 8, 2007) supports the research team's analysis of her mathematical understanding:

> [Vena] used a guess-and-check process to build a comparable figure [see Fig. 5] to the last one (*Task 1f*) in the manual. We videotaped her while she attempted to make sense of the connections. She eventually made the connection through moving the [2-D conventional on-screen] figure to more closely match the orientation in the control panel [see Fig. 6].

Several weeks later, during Lesson 20 (February 27, 2008), this objective was revisited, since the research team wanted to determine the extent to which the children had mastered top-view coding when given 2-D conventional pictures. Although the children continued to develop codings for unusual combinations of Soma figures, for this mid-program checkup, the team revisited Task 1f in the manual (see Fig. 5). Each child drew a new, blank grid on a sticky note to cover the manual's printed grid that had previously been filled in, and then wrote in the numbers on the sticky-note

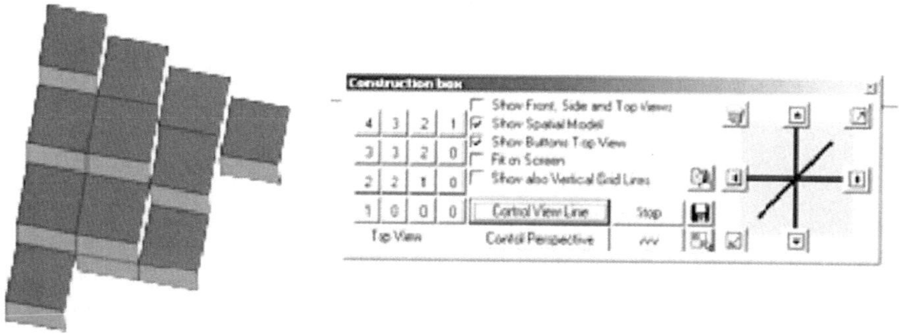

Fig. 6 The dynamically rotated figure closely aligned with the top-view coding grid

grid based on how he or she now viewed the figure. Next, each child checked his or her prediction by building the figure on the Geocadabra Construction Box using the new sticky-note numbers. However, before entering the numbers into the Construction Box' four-by-four grid, the children clicked the "Hide spatial model" option to prevent them from seeing their structures emerge as they entered the grid numbers. After entering all the numbers each child self-checked the computer-generated figure against the picture in the manual by un-clicking the Hide option to show the completed figure on the screen. The research team paid particular attention to Vena who now performed this task flawlessly. Over time she appeared to have developed the competence that had not been observed a few weeks before. They concluded that learners develop these particular skills at different rates regardless of the number activities they experience.

Lesson study components, in particular, (1) consider goals for student learning and development; (2) plan a research lesson based on these goals; (3) during the research lesson, collect data on student learning and development; (4) use these data to reflect on the lesson and on instruction more broadly, and (5) if desired, revise and reteach the research lesson to a new group of students, are evident in these example lessons as the research team revises the manual and other activities for new groups of students who will experience the program in the future. The research team decided to introduce the Construction Box activities earlier and the team will check for top-view coding competency more carefully with future cohorts of participants.

Example 3: Children's Use and Invention of Top-View Coding to Represent 2-D Figures

This example illustrates how the research team modified multi-lesson activities based on their close attention to student understanding and consensus that the activities should be differentiated to meet individual needs. By January 2008, the

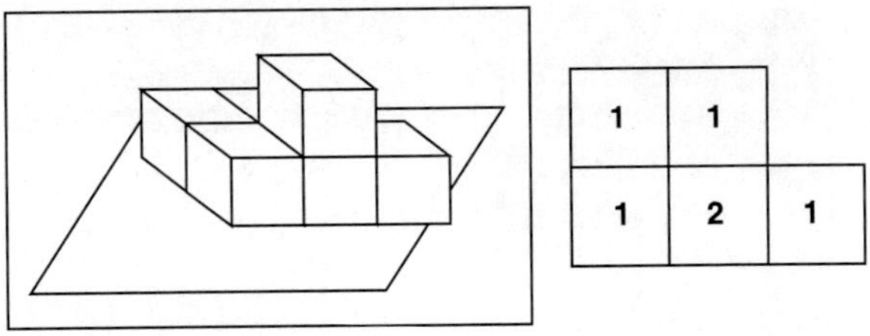

Fig. 7 Two-figure Soma assembly task card and the corresponding top-view coding

children had mastered the objective to produce numeric top-view codings (abstract representation) of many of the figures (2-D representation) on the task cards. For an example, see Fig. 7.

Lessons were designed to be relatively open ended with self-selecting levels of challenge and interest. The children, in pairs or independently, first selected a task card, then identified the two Soma figures that comprised the illustrated figure and drew the numeric top-view coding to represent the figure, and finally, traded the coding with another student or pair of students who in turn tried to re-build that figure directly from its student-generated coding. Some children were able to identify the two Soma figures directly from the assembly coding using mental imaging while others used Soma figures or even loose cubes to check their solutions. In general, both classes handled this activity very well and continued to be enthusiastic about creating coding puzzles for each other to solve. At this time, the research team was unaware of a conventional coding for figures with holes or overhangs such as those that appeared on several of the assembly task cards. For an example, see Fig. 8. They were curious to see how the children would code such figures. This is an instance where children's inventions were honored.

On December 11, 2007 (Lesson 12), the research team noted that there were two different but mathematically appropriate ways to interpret the numbers in the conventional coding system. Reflection notes show that to some children, "'3' meant a stack of three cubes on a space, while others understood '3' to mean a cube on the

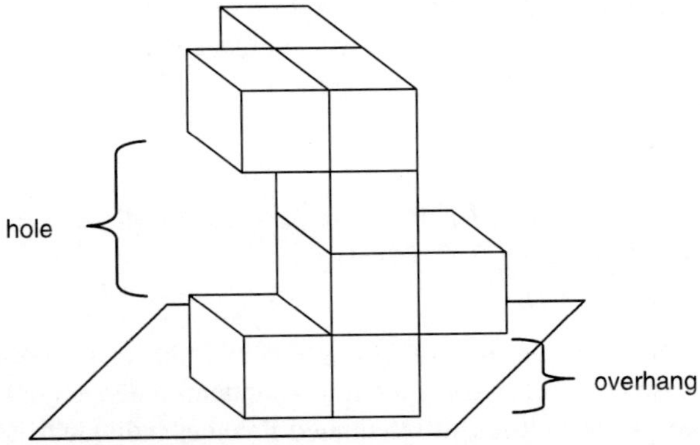

Fig. 8 A two-figure Soma assembly with a hole and an overhang

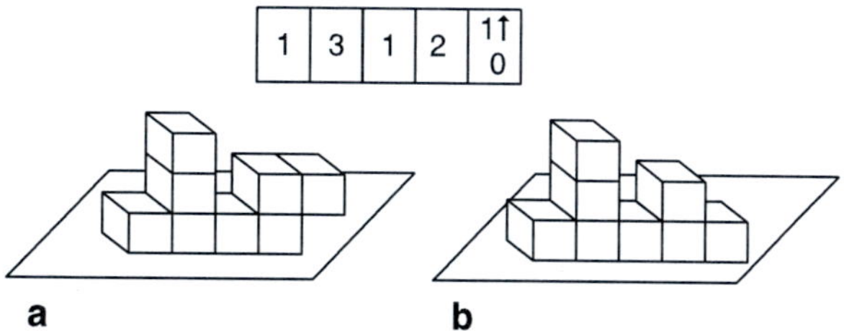

Fig. 9 Abby and Belinda's challenge. **a** Abby and Belinda's coding and figure. **b** Megan and Gina's interpretation

third level of a stack of cubes. Both interpretations are accurate." Confusion arose when these two ways of interpreting numbers on the grid came into conflict when given the challenging task of creating their own coding system for figures that had holes or overhangs. Based on reflection notes for Lesson 12, the research team noted: "One pair of students coded an overhang using '**3' to mean two spaces (**) below a cube on the third level. Others used '1↑↑' to mean one cube raised above two spaces."

When the program resumed after the winter holidays, the coding invention challenge continued. Based on notes from Lesson 13 (January 9, 2008), Fig. 9 illustrates how Megan and Gina jointly misinterpreted Abby and Belinda's coding. Figure 10 shows Kevin's figure and coding and the coding John suggested to correct it.

The plethora of invented codings created confusion. To remedy this confusion, Irma called a class meeting to discuss the need for a class-wide coding system that would continue to honor the conventional system but would also include a way to denote holes or overhangs. She invited them to share or create ways to solve this dilemma. Using Sarah's task card and a 3-D model of this figure, each student drew his or her own way of coding the spaces on the left side of the figure. Irma con-

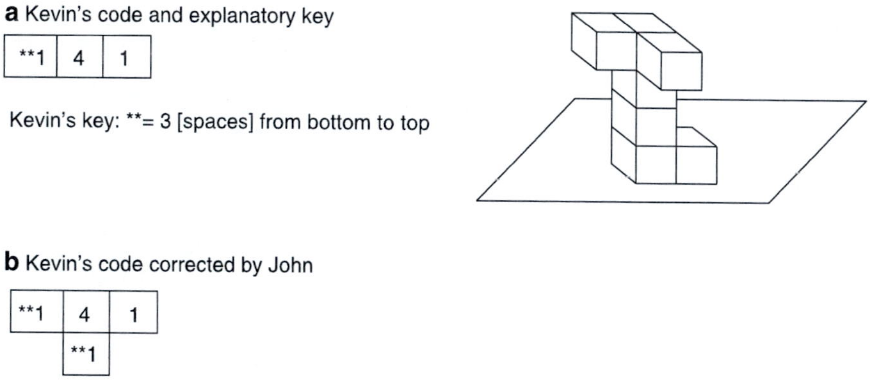

Fig. 10 **a** Kevin's figure and coding. **b** Kevin's code corrected by John

Fig. 11 Sarah's task card and coding

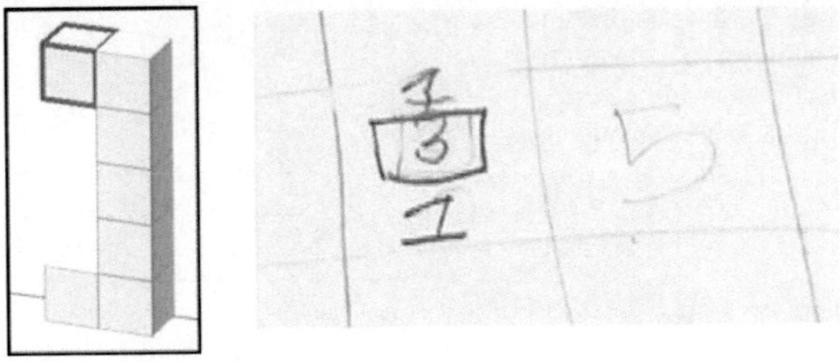

ducted a negotiation discussion, during which each child who wished to present a coding system was videotaped while the others attended respectfully. Student work is shown in Figs. 11–15.

Sarah:	Right here it has 3 spaces, 1, 2, 3 [Pointing to the gap on the left side of the figure]. Right here on my paper, I put a 3 in the middle, 1 on the top and 1 on the bottom, Because there's a 1 right here [pointing to the 1 above the space on the 3D figure]; there's another 1 right here [pointing to the 1 below the space on the 3D figure]. There's 3 spaces.
Jackie:	The 3 inside the square on your code represents spaces?
Sarah:	Yes.
Jackie:	Number of spaces?
Sarah:	Yes.
…	
Maddie:	A 1 here (pause) and then I put (pause) and then to the side I put a rectangle with, with 3 squares in it. And the key is right here.
Jackie:	Those 3 squares mean 3 . . . (pause)
Maddie:	Cubes.
Jackie:	I don't see 3 cubes there.
Maddie:	I mean 3 cubes . . . (pause)
Irma:	Missing?
Maddie:	Yeah.
Jackie:	3 blank spaces?

Fig. 12 Maddie's coding

Fig. 13 Maddie's key

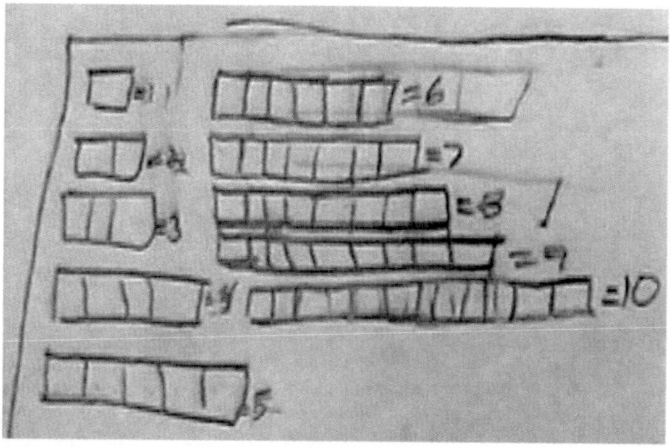

Fig. 14 Gary's coding

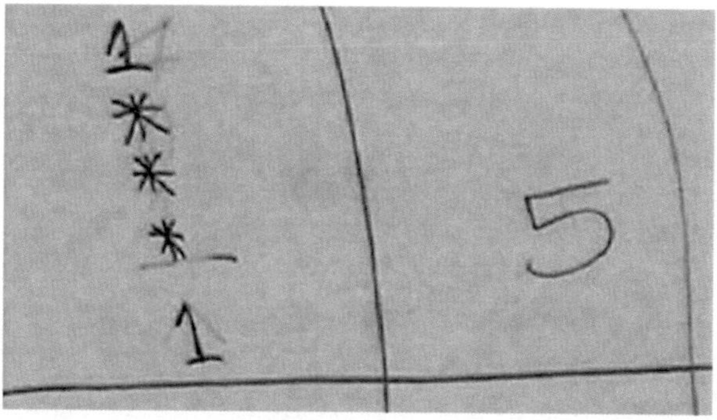

Maddie:	Yeah.
Jackie:	Where there could be 3 cubes but there are 3 blank spaces.
…	
Gary:	I have 1 right here 'cause there'll be 1 at the bottom over here [pointing to the 1 cube below the space on the 3D figure]. I have 3 stars (pause) for (pause) there's 3 (pause). Each star represents 1 space. So that will be 3 spaces just like here: 1, 2, 3 [pointing to the 3 stars on the 3D figure]. Then I put 1 on top 'cause there's 1 on top of the spaces. And I put a 5 right here because there'll be 5 going up – 1, 2, 3, 4, 5 [pointing to the 5-cube stack on the right of the 3D figure].
…	
Irma:	I have a question for all of you. Some of you have used stars; some of you have used a box with a number; some of you have used 3 squares to show the same structure. How are these 3 codings alike?
Elliot:	They all mean 3 spaces.
Irma:	I don't understand, "They all mean 3 spaces."
Elliot:	They all mean that 1 is on the bottom, and there are 3 spaces and 1 on the top.
Irma:	Which one will be the best way to code?
Elliot:	His! [pointing at Gary's coding]—because he has 3 stars.
Irma:	So instead of stars, how is that different from Maddie's? Maddie has 3 boxes. What makes it better to have 3 stars?
Gary:	'Cause, if you put 3 boxes, they might say that there's 3 stacking.
Irma:	Oh, 3 stacking on each other.

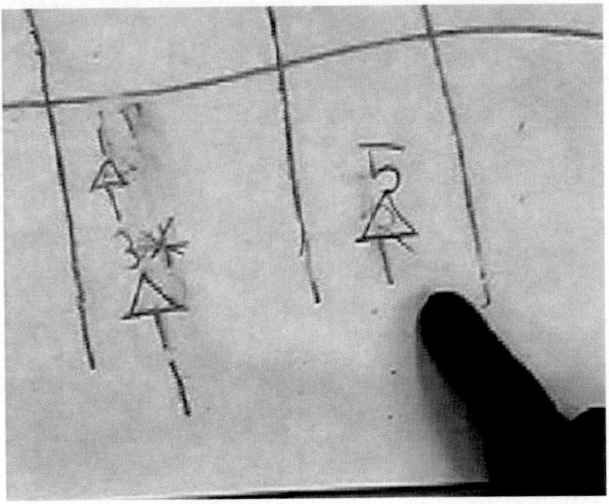

Fig. 15 Elliot's coding

Sarah:	With mine, because it's easier and it's not that long.
Irma:	Oh, what do you mean, it's not that long?
Sarah:	Because it has only 4 steps.
Irma:	Are you telling me that if I want to change the number of empty spaces, all I have to change is the number inside that box [pointing to the square with the 3 on Sarah's coding]?
Sarah:	Yes!
Irma:	What would Gary have to do?
Elliot:	Make more stars.
Irma:	Now let's think about Elliot's.
Sarah:	Ooh! Ooh! I've got a question for both of them!
Irma:	We're going to wait for your question. I have a question for all of you.
Elliot:	I can do the same thing that she can do.
Irma:	Tell me.
Elliot:	I already have the space and I put the numbers.
Irma:	So yours and hers are alike.
Elliot:	Because, both of us, all we have to do is change this number.
Jackie:	Does that mean Gary's and Maddie's are alike because they're showing how many [spaces] by 1s? Stars here and squares here.

The class was invited to choose from among the presented coding systems and each student was invited to defend his or her choice. They agreed to use 3rd grader Sarah's coding system with a class-negotiated modification: the use of a circle instead of a square to denote the number of empty spaces. This open-ended task presented students with an opportunity to be creative, to evaluate each other's work respectfully, and to come to consensus over a new convention that all would use.

This example illuminates how the research team's planning focused on providing a highly differentiated learning environment in which students select from a range of options. In both classes children had the freedom to work alone or with a partner. In selecting task cards, they had different levels of complexity from which to choose. Some chose figures with holes or overhangs, others chose figures that did not clearly identify the two Soma pieces from which they were formed. Children who lacked confidence in coding complex figures selected cards with simpler figures that were easy to code and clearly distinguished the two individual Soma pieces. Furthermore, the negotiation activity provided students the opportunity to present or to critique each other's invented coding systems. The lesson study components, (1) consider goals for student learning and development; (2) plan research lessons based on these goals; (3) during the research lesson, collect data on student learning and development; (4) use these data to reflect on the lesson and on instruction more broadly, and (5) if desired, revise and reteach the research lesson to a new group of students, are all evident in these lessons. Of note, during the research team's weekly reflection meetings, discussion centered around whether students attained intended learning objectives, based on observation, informal interviews, and hard data from the lesson. Then the team negotiated what should come next. The learning trajectories emerged in a highly constructivist manner, based on the team's perceptions of student interest but always with the SOC representations as a guide.

Conclusion

The inquiry-based characteristics of lesson study correlate with the research team's design research process. The inquiry process is fluid and ever changing (Craig 2008), and "unfurls in a non-linear fashion" (p. 1995). Reflection on practice, an essential component in any inquiry-based process, is at the heart of good instruction and at the heart of good research. Figure 16 illustrates this research team's view of how lesson study and design research intersect.

Considering the Lesson Study Activities of teaching the research lesson, the team collects, analyzes, and reflects on student learning, and sometimes revises during and after every planned activity. This is demonstrated in Example 1 when students in the 3rd-grade group were able to bypass intended learning objectives by selecting alternate solution pathways that the team did not deem to be productive for that lesson's focus.

For the Teacher-Researcher Learning component, the focus on student understanding has impacted the knowledge base in the domain of 3-D visualization. Having their research validated when they presented some of their findings at an international research conference (Sack and Vazquez 2008) has raised the research team's sense of efficacy as new researchers in the field of curriculum development. The research team's attention to detailed field notes, coupled with transcriptions of key video clips and copies of students' written work provides triangulated evidence about how students respond to the developing activities. The ongoing study, now in Year 3, continues to evolve. Activities from Year 1 have been used and revised with larger groups of 3rd-grade children (approximately 20 children attended consistently in Year 2, and 28 in Year 3). Each lesson and its objectives are adapted to meet the needs of each group of students. This is a core component of lesson study. For example, during Year 1, classes were based in Irma's classroom using the school's

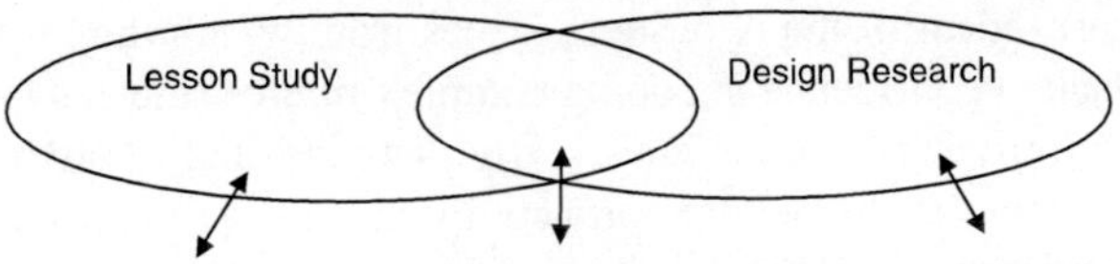

Lesson Study Activities	• Observe the research lesson and collect data on student learning and development • Use the data to reflect on instruction more broadly	• Plan a research lesson based on goals for student learning and development • If desired, revise and reteach the research lesson to a new group of students	• Teach the research lesson and collect data on student learning and development • Use the data to reflect on the lesson and student learning
Teacher/ Researcher Learning	• Motivation	• Increased knowledge and connections among knowledge • Sense of efficacy • Professionalism • Lesson plans • Representations/models • Manipulatives/worksheets • Assessment protocol	• Dissemination of knowledge
Outcomes		• Improved classroom practice	• Curricular resources for the development of spatial skills

Fig. 16 Lesson study and design research Venn diagram

portable laptop system when needed. Much larger Year 2 and Year 3 classes require the use of the school's computer lab. Many activities are now conducted on the carpeted floor of the lab where there are no group tables as in Irma's classroom. Making adjustments for class size and room organization has been a necessary but important challenge to the research team as they consider flexibility in instructional settings as part of the program's development.

Although an immediate priority is to develop knowledge about how students interact with the intervention activities, the professional development of each team member cannot be understated. The lesson study components of planning lessons based on SOC-guided goals, developing materials for these lessons, paying close attention to student understanding during the lessons, and debriefing to revise and plan ongoing lessons have become a natural part of each member's weekly routine and professional life. Team members are able to frankly discuss each other's instructional practices. For example, one member tended to call on the same children when asking the class for input. The other team members felt that other, less-confident children should be given opportunities to shine and to develop self-efficacy in front of the class. Thus, with respect to the lesson study Outcomes component, these conversations and subsequent practices in the research classroom have impacted the general classrooms of all three team members.

When the project began in 2007, the research team believed that they might have a set of publishable lesson plans by the end of the second or third year. Now in the middle of Year 3, they realize that their lesson plans continue to be refined and evolve according to the needs of each group of students. At this time, plans are still "under construction" and are only available to teachers who attend professional development sessions led by research team members. The team envisions going through additional iterations before they can begin to formalize the currently used lesson plans.

As they continue to teach, observe, and reflect on their research lessons, the research team wonders about the extent to which individual or group characteristics influence their understanding of children's sense making of spatial concepts. This question remains at the forefront of their reflections as they find themselves moving in new directions with each new group of 3rd graders in their study.

References

Borba, M. C., & Villarreal, M. E. (2005). *Humans-with-media and the reorganization of mathematical thinking*. New York: Springer.

Cobb, P., Confrey, J., diSessa, A., Lehrer, R., & Schauble, L. (2003). Design experiments in educational research. *Educational Researcher, 32*(1), 9–13.

Cochrane-Smith, M. (2003). Learning and unlearning: The education of teacher educators. *Teaching and Teacher Education, 19*(1), 5–28.

Craig, C. J. (2008). Joseph Schwab, self-study of teaching and teacher education practices proponent? A personal perspective. *Teaching and Teacher Education, 24*(8), 1993–2001.

Dreyfus, T. (1991). On the status of visual reasoning in mathematics and mathematics education. In F. Furinghetti (Ed.), *Proceedings of the 15th International Group for the Psychology of Mathematics Education Conference* (Vol. 1, pp. 33–48). Genova: Univ. de Genova.

Freudenthal, H. (1991). *Revisiting mathematics education: China Lectures*. Dordrecht: Kluwer.

Lecluse, A. (2005). Geocadabra [Computer software]. Retrieved July 28, 2007, from http://home.casema.nl/alecluse/setupeng.exe

National Council of Teachers of Mathematics (NCTM). (2000). *Principles and standards for school mathematics*. Reston: Author.

National Research Council (NRC). (2006). *Learning to think spatially: GIS as a support system in the K-12 curriculum*. Washington: Author.

Presmeg, N. C. (1992). Metaphors, metonymies and imaginative reality in high school mathematics. *Educational Studies in Mathematics, 23*(6), 595–610.

Sack, J., & van Niekerk, R. (2009). Developing the spatial operational capacity of young children using wooden cubes and dynamic simulation software. In T. Craine & R. Rubenstein (Eds.), *Understanding Geometry for a Changing World: Seventy-first yearbook* (pp. 141–154). Reston: NCTM.

Sack, J., & Vazquez, I. (2008). Three-dimensional visualization: Children's non-conventional verbal representations. In O. Figueras, J. L. Cortina, S. Alatorre, T. Rojano, & A. Sepulveda (Eds.). *Proceedings of the Joint Meeting of International Group for the Psychology of Mathematics Education 32 and PME-NA XXX* (Vol. 4, pp. 217–224). Mexico: Cinvestav-UMSNH.

Tall, D., Gray, E., Bin Ali, M., Crowley, L., DeMarois, P., McGowen, M., Pitta, D., Pinto, M., Thomas, M., & Yusof, Y. (2001). Symbols and the bifurcation between procedural and conceptual thinking. *Canadian Journal of Science, Mathematics and Technology Education, 1*(1), 81–84.

Tomlinson, C. A., & McTighe, J. (2006). *Integrating differentiated instruction and understanding by design: Connecting content and kids.* Alexandria: Association for Supervision and Curriculum Development.

van Niekerk, (Retha) H. M. (1997). *A subject didactical analysis of the development of the spatial knowledge of young children through a problem-centered approach to mathematics teaching and learning.* Unpublished doctoral dissertation, Potchefstroom University for CHE, South Africa.

van Niekerk, R. (2006). Build and explore with Geocadabra (J. J. Sack, Trans.) Retrieved July 28, 2007, from http://nsfmli.rice.edu/curriculum/Lecluse/GeocadabraGeobuddyConstructionBox-EnglishTranslation.pdf

Yakimanskaya, I. S. (1991). The development of spatial thinking in school children. In P. S. Wilson & E. J. Davis (Eds. & Trans.), *Soviet Studies in Mathematics Education* (Vol. 5). Reston: NCTM.

Lesson Study: A Case of the Investigations Mathematics Curriculum with Practicing Teachers at Fifth Grade

Penina Kamina and Patricia Tinto

Background

Practicing 5th-grade teachers find themselves often at the heart of reform initiatives with inadequate professional development or support. Teachers are expected to shed their old pedagogical beliefs, comprehend mathematical content in new ways, and effectively implement instructional materials such as *Investigations in Number, Data, and Space* (*Investigations*) (Putnam 2003) in their mathematics classrooms. In one such instance, the discrepancy between the teachers' prior experiences, National Council of Teachers of Mathematics [NCTM] (2000) principles, and *Investigations* objectives presented an important problem for study. The conceptual framework of this study is that classroom-based instruction of school mathematics is a negotiation between students, teachers, and curricular materials with the teacher as initiator of what actually takes place in the classroom. This case study research looks specifically at three teachers who collaborated with each other in lesson study, and describes their attempts to establish new classroom instructional approaches as they facilitated *Investigations* lessons.

The emerging in-depth competency in teacher content knowledge and teacher pedagogical knowledge, and changes in teacher pedagogical beliefs developed from the collaborative process of lesson study enabled these teachers to rethink their practice. One teacher implemented new instructional strategies; the second teacher emerged as the mathematics coach and collaborative leader for the group, and the third teacher, while not yet able to put new strategies into her classroom instructional routine, did gain the ability to reflect on her practice and to share those reflections with others.

P. Kamina (✉)
SUNY College at Oneonta, Oneonta, NY, USA
e-mail: kaminapa@oneonta.edu

L. C. Hart et al. (eds.), *Lesson Study Research and Practice in Mathematics Education*,
DOI 10.1007/978-90-481-9941-9_17,

Theoretical Background

Two important yet little understood questions guiding this research are: (1) How do teachers negotiate the demands of new curriculum materials within their established instructional practice and their own pedagogical beliefs? (2) What are the competing factors that teachers have to negotiate to facilitate standards-based curricula? Several mathematics studies at all school levels profess that students who use these standards-based materials attain higher test scores on local examinations, as well as on measures of conceptual understanding (Cramer et al. 2002; Flower 1998; Goodrow 1998a; Mokros 2000, 2003; Reys et al. 2003). These studies are concerned with the effects of curriculum on student achievement but fail to enlighten practicing teachers about the classroom cultures that existed for these achieving students. This lack of comprehension about reform classroom cultures is primarily curriculum-dependent. Again, the standards-based materials are facilitated by practicing teachers who have taught using teacher-centered methods without benefit of professional development on inquiry-based instruction (Fernandez 2005; Kamina 2005).

As such, this study views the instruction of mathematics as the negotiation of practices of school mathematics with the teacher as initiator. For the purpose of this study, negotiation involves reasoning, interpreting, and making sense of mathematical meanings presented by the students (Frid 1994). A teacher's subject matter knowledge and pedagogical content knowledge (Shulman 1987; Ball and Bass 2000) are also instrumental. More recent studies have expanded the idea of the knowledge needed by a teacher to categories described by Ball et al. (2008) as "horizon content knowledge, common content knowledge, specialized content knowledge, knowledge of content and students, knowledge of content and teaching, [and] knowledge of content and curriculum" (p. 403).

Context of Lesson Study

This lesson study case focusing on 5th-grade teachers in an urban school district was part of a larger project that explored the facilitation of the *Investigations in Number, Data and Space* (TERC 1998) curriculum. Data were collected during the 2003–2004 academic year for the larger study from all 5th-grade classrooms at Lake (pseudonym) Elementary School. The first author was conducting her dissertation study (Kamina 2005) and was an observer during lesson study meetings and the second author had given professional development workshop at this site in the 2002–2003 academic year.

Lake is in a midsize city school district in the northeast United States and was suited to this study because of its adaptation of *Investigations* as its primary curricula. All the teachers at Lake used *Investigations* materials. Additionally, the school focused on instructional issues, a key aspect of this study. All teachers met regularly in grade level team meetings that concentrated on a topic for professional development using the *Investigations* curriculum.

Moreover, the 5th-grade team committed themselves to developing as a learning community to grow professionally through lesson study that also had the school administrators' support. Lake school administrators, when analyzing State assessment results in Mathematics and in English Language Arts, noticed a performance and achievement difference at each grade level. At each grade level there was one class that did significantly better in the State assessments and outperformed the other two classes. Thus, the administrators wanted the lesson study conducted in order for all their teachers to have an opportunity to explore what leads to exemplary practice within the school.

Investigations is a K-5 standards-based curriculum with nine units at 5th grade, whose objectives offer students connected and meaningful mathematical problems to promote in-depth thinking. *Investigations* is designed to develop students' conceptual knowledge and it encourages teacher use of inquiry-based instruction as broadly defined. The study's participants addressed mathematical concepts in two of the nine units during the lesson study cycle. These were the *Name That Portion* unit that deals with the equivalence and how fractions, percents, and decimals are related, and the *Picturing Polygons* unit that deals with two-dimensional geometry and the relationship between side lengths and areas of similar shapes.

All the 5th-grade classrooms at Lake had student ages ranging from 10 to 12 years with relatively the same distribution of students in total number, gender, and academic abilities. However, the first class, an inclusive class, had two female teachers for the academic year that shared instruction. Teacher Zsa, the elementary certified lead teacher, taught the mathematics lessons and teacher Ivey, who held certification in Special Education, led the class during English language lessons. Both teachers actively participated in the lesson study cycles. Besides these two teachers, there was also a teaching assistant. In this inclusive class, there were a total of 25 learners, 11 girls and 14 boys. Nine out of the 25 students were students with documented special needs. The 2003–2004 academic school year was Zsa's second year of both teaching 5th grade and using the *Investigations* mathematics curriculum units. It was Ivey's first year as a 5th-grade teacher, having moved from 4th grade.

Students in the second class had two female teachers, Suzie and Nora, in succession. Suzie briefly taught the second class at the beginning of the school year before departing for maternity leave. Nora took over as a new teacher with only student teaching as a prior teaching experience. This was Nora's first experience using a standards-based curriculum. There were 22 students in this class of 13 boys and 9 girls with no teaching assistant during mathematics lessons.

A male teacher, Jesse, taught the third class. His class had a total of 26 learners, 15 girls and 11 boys and no teaching assistant. Jesse noted that he had used *Investigations* for the last eight years, in 4th and 5th grade, beginning with its pilot at Lake Elementary until its adoption in 2003 as the school district's K-5 mathematics curriculum. He was facilitating the 5th-grade *Investigations* units for the fifth time. Jesse was the only teacher with previous knowledge of lesson study. He attended a professional development workshop on lesson study in the summer 2003.

Method of Data Collection

A qualitative case study research design was used to explore the teachers' emerging practice models in facilitating *Investigations* mathematics in the three urban 5th-grade classes. Data were collected through interviews, classroom artifacts, video of three lesson study planning and debriefing meetings, audio of lesson study teaching, audio and notes of weekly 5th-grade team meetings, mathematics lesson plans for the 2003–2004 academic year, several classroom lesson observations, and pre- and post-lesson interviews. Once a month, a 30-minute in-depth interview was carried out with each of the classroom teachers, in which they answered open-ended questions arising from the classroom observation notes. All field notes and observations were transcribed and analyzed to reflect the teachers' voices and narratives.

Analytic induction was used in coding and data analysis (Bogdan and Biklen 2003). For example, the first five classroom observations were analyzed by coding field notes, listening, and viewing the recordings for recurrent themes and patterns. We further pursued those themes that emerged in-depth by focusing on these categories during further data collection. This helped to narrow down on issues pertaining to curriculum and inquiry-based instruction. Post-lesson interviews were held with each teacher to understand their perceptions of the *Investigations* curriculum. Questions such as, "How would you teach this lesson next time?" and, "Tell me about the lesson," were posed.

Lesson study data were collected over five months beginning in November, when the classroom's sociomathematical norms and routines had been set and the teachers had increased the classroom time allotted to mathematics instruction. Three lessons were planned using the lesson study design as modified by the team. Only two were taught due to the school closing for severe weather. The team identified the lead teacher for each lesson study cycle who then drafted a lesson plan on an agreed-upon math concept. Copies of the drafted lesson plans were then given to each of the team members to be discussed at their meeting. The lesson study meetings were held during school hours, but substitute teachers were used in the classrooms of the teachers who were observing and for all classrooms during debriefing meetings.

Results of the Lesson Study

The first focus lesson was based on the curriculum unit *Mathematical Thinking at Grade 5*. Jesse, the most experienced teacher, volunteered to be observed. The lesson study planning discussions concentrated on content, the classroom climate, and the students' work, on how to introduce the lesson and on how to facilitate the students learning the specified mathematics concepts:

Jesse: …so we can review materials like that finishing both division clusters, section five factor pairs of 1,100 and get into Investigation Four which really is a lot of groups of 10,000's stuff, and they [the students] seem pretty good at that game already, cluster thousands.

Ivey: We played that a lot in fourth grade.

Jesse: A thousand, or a hundred?
Zsa: A hundred.

During the lesson implementation, Ivey, Zsa, and Nora observed the teaching from seated positions located in different parts of the class. Jesse launched the lesson by asking students what math they had been working on lately, which elicited several responses that were, written on the board:

Jesse: …factors. What two numbers, when multiplied, equal another number. Those are the things we talked about. Like on our chart, when we said, "what are factors," "what are multiples," "list the factors of 20." And you told me: 1, 2, 4, 5, 10, and 20. Those are factors. Multiples of 20, though, are something else. What are the multiples of 20? …20…40…60…
Student: 80, 100, 120, 140…
Student: 160…
Jesse: Alright. Those are multiples. We need to make sure—or I should say, you need to make sure—that when we talk about multiples and factors, you understand which is which. Sometimes they're easily mixed up. Alright. We're going to transition into the rest of our math activity today. And for this, this is what I need you to do. Now listen to all these directions. They're multi-step directions.

Next, he displayed a poster chart that had on it "Finding factor pairs of 1,100." Below this title were the instructions, "(1) Find some factor pairs of 1,100 and (2) Write about how you found each factor pair." Jesse asked the students to take out their math journals and work in small groups. He moved around the room as students worked in their groups. He heard a group saying, "You have to write it as a factor pair," and questioning, "What does a pair mean?"

Twenty minutes later, Jesse held a whole-class discussion on their findings. He wrote students' solutions and strategies on the chalkboard and on a chart. He also had the students discuss these items by posing probing questions.

At the end of the lesson, a substitute teacher stepped into the observed classroom allowing the lesson study group to debrief in another room. The group noted several successful instructional strategies including details that were not discussed during the planning session. For example, Jesse had created and posted a chart on how students were to work in a small group, and he pinned classroom artifacts and Math vocabulary cards on or around the chalkboard. Each vocabulary card was made from different colored construction paper and was pinned above the chalkboard where all students could refer to it. The vocabulary cards had words such as "patterns," "rectangle," "factor pair," "prime number," "multiple," "dimension," "square number," "skip counting," and "the power of 10." Jesse also displayed a chart of factor pairs of 100 and of 1,000 as well as a direction chart that had the following instructions on it:

When I think I'm "done" with my Math, I will:

- Check it with my group
- Look for a way to do extra or extend the work
- Explain my thinking in my math journal—my strategies, the tools I used
- Look for another way to arrange the numbers

Figure 1 shows some of the artifacts that supported the first lesson study as well as evidence of students' work: the *Mathematically Speaking* notice board with the

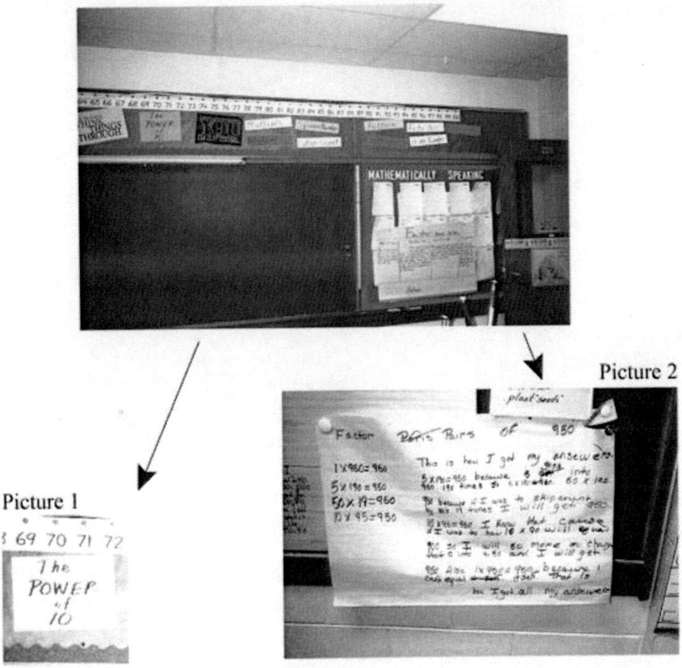

Fig. 1 Notice Board. *Picture 1* The Power of 10. *Picture 2* Factor Pairs of 950

current terms being used in the class; an orange colored card that reads "the Power of 10" with energy emanating out of 10 (see Picture 1); Jesse's work and a student's work on factor pairs of 950 (see Picture 2). In Picture 2 the student vertically lists the factors pairs as "*1 × 950 = 950, 5 × 190 = 950, 50 × 19 = 950, 10 × 95 = 950*" and justifies reasoning in writing: "*This is how I got my answers. 5 × 190 = 950 because 5 goes into 950, 190 times so 5 × 190 = 950. 50 × 19 = 950 because if I was to skip count by 50's 19 times I will get 950. 10 × 95 = 950 I know that cause if I was to have 10 × 90 will equal 900 so I will 50 more so change 0 into a 50 and I will get 950. Also 1 × 950 = 950 because 1 can't equal itself. That is how I got my answers*."

Jesse not only used the *Investigations* materials as provided by its developers, but he also demonstrated specialized content knowledge, knowledge of content and students, knowledge of content and teaching (Ball et al. 2008) as well as a positive pedagogical disposition towards *Investigations*. The teachers held an enthusiastic debriefing discussion about this first lesson that went beyond their planning. The teachers talked about how Jesse involved the students from the introduction of the lesson to its closure, the varied forms of ideas and questions that permeated the lesson, the structure of the lesson, the organization of the class, and the management strategies he had in place. The teachers in their collaborative dialogue even distinguished between "powers of ten" and "multiplying by ten—adding zeroes," while clarifying a misconception held by Zsa. She was now clear in differentiating "powers of ten" as adding the numbers of zeroes corresponding to the exponent of ten to a given number and "multiplying by ten" as adding only one zero to the number to be multiplied.

The second focus lesson on the unit, *Name that Portion* was planned but not taught due to the school closing for severe weather. When substitute teachers became available, all teachers had finished teaching the focus lesson and needed to select another unit for observation. However, the first author of this chapter had observed each of the three teachers implementing the planned lesson. During the planning session, the teachers had decided that simultaneously introducing fractions, decimals, and percent simply by posing a brainstorming question to students asking them to name everyday uses of the three concepts and then discussing the students' responses would be a very valuable strategy. Both Jesse and Zsa posed the question, "Tell me where you see or use fractions, percents, and decimals in everyday life." An in-depth discussion following the brainstorming of responses such as, "percents in class grades; decimals in money; fractions of pizza at a party," took place.

Nora, however, began this lesson by restating classroom management expectations and reviewed the multiple representations of part of a whole.

> Nora: Take out your math book, math journal, and pencil. Once math starts, nobody is talking, all right. Ten, nine, eight... one [*Teacher counts down to obtain students' attention and for silence*] everybody in this front row is ready to go. This is what I would like you to draw. On a clean page in your journal with the date on top [*students talking*]—Guys, can I do this? And, can you please quietly fill out your journal? This should look familiar to you. What does this remind you of?

Nora taught using a well-established routine. She guided and led the students each step of the lesson. Nora did not follow the plans for a brainstorming session but rather asked the students to come up with equivalent fractions, decimals, and percents independently:

> Nora: Please take a minute to think of an example where you can make an equivalent all the way across, from a fraction, a percent, and a decimal.... Do this quietly, by yourself. ... Okay, listen to the directions again.

Table 1 was written on the chalkboard.

She circulated among the students observing that the students stayed on task and the table was filled in correctly. She then led a whole-class discussion about the addition of two fractions using the least common multiple approach.

Zsa led the third focus lesson on the unit *Picturing Polygons* that deals with two-dimensional geometric concepts of polygons such as identifying and sorting polygons, finding angle sizes, turns, and similarity. During the three-hour planning session, several issues relating to the demands of facilitating a new curriculum unit were discussed.

The teachers first discussed prerequisites that students must have been introduced to in lower grade levels. For example, the teachers questioned whether

Table 1 Different Representations written by Nora

Fractions	Percents	Decimals
1/1	100%	1.00

students might have learned some basics of the content they were planning, and anticipated challenges in the absence of prerequisite knowledge.

Jesse: Do you think you will actually get to use the Power Polygons?
Zsa: Well, I think I'm going to start by introducing it by seeing if they remember the three different kinds of angles you can have. And, then you know, just do the kind of quick review of what a cube is and the….
Ivey: Some of my kids are like, "We never used these before." And we were like, "If you were in my third grade class, I know you used them." The first year I taught third grade, I taught it for a week because I thought I had time left….

By this lesson, the teachers were developing a "curriculum lens." Their discussions were going beyond the specific lesson to look at student learning across grade levels and they were questioning where students began to learn needed concepts (Fernandez et al. 2003). Teachers, however, did not attempt to change the course materials. Thus, the materials defined the lesson at this point and provided the mathematics content. Conversations from the planning session highlight the teachers' mathematics content knowledge and their prior math attitudes. Zsa, for example, stated that as a student she disliked geometry topics such as scalene and isosceles.

Ivey: We all see things differently but if we do not even know what we are talking about as instructors, which is another issue that I think that is difficult. Up until four or five years ago, I had to teach myself scalene.
Zsa: Because, after high school, that was it. I did not ever want to deal with angles again.

Zsa and Ivey also had concerns about the use of supplemental materials to the *Investigations* lesson that might be more appropriate to use with some of the students in their inclusive classroom.

Ivey: Do you think I need to bring in more outside information? I'm still at this stage now where I'm still getting used to the *Investigations,* and I am still sticking close to what the book is saying to do. Should I be bringing in more outside examples or materials, to facilitate them [students] getting this?

The teachers reached an agreement to stay focused with the *Investigations* materials and not to bring in supplemental materials and examples.

Near the end of the planning session, the teachers discussed what should be observed.

Jesse: What do you want us to look at?
Zsa: Well, this is the lesson where I think I will have the questioning. Make sure my questioning is what it needs to be. You know my wording. Especially, am I skipping that one little component part?

Zsa was concerned that her teaching strategy was not student-centered and was making an effort to improve. Thus questioning was the strategy she wanted to document. Jesse and Ivey also proposed that Zsa should watch and listen to student interactions especially to what students might be saying to their partners or groups, and listen to students' use of mathematical terminology and language. Important to the focus on questioning was how the teachers chose to group the students for interaction, discourse, and cooperative work:

Ivey: Probably, in pairs. Well, I usually say between two and four kids, and I let them choose their own groups between two and four. Then, I do not know. Is four too big for this stuff?
Jesse: Uhm, I think four would turn into two twosomes anyway.
Ivey: You know what, you are right. That rather tends to happen anyway.
Jesse: Which then they can use each other and to check with or discuss with.

Besides, the aspects of planning cited above, the teachers also looked at the connections among the *Investigations* curriculum units. The lesson study planning prompted the teachers to consider coherence of the curriculum. The teacher who had previously used the curriculum suggested ways for the 'newer' teachers to bridge topics. Thus, the lesson study process facilitated the sharing of expertise from the experienced teacher to the less experienced teachers.

Finally, the third lesson was implemented and debriefed. Below is an excerpt from this meeting:

Zsa: Well I know I lost them at some point. Definitely when we went on to the second work showing using the circle, the center point because I know I did not explain myself very well when I started so they didn't know they were only supposed to use the small angle. So, there was that and I don't know if I spent too much time doing it.
Jesse: On the center point examples?
Zsa: ...I did not see anybody using the center point, but see; now I am thinking I do not think I explained it very well for them to understand....
Jesse: And I think you can actually say, you use the pieces, you have to be combining and putting together as we did with the red triangle, you have to put pieces together. ... You are going to use your known, what is with the green equilateral. I mean, I think at this point most of them know that, so have someone bring that one up so I know this 60, 60, 60, and I can use this and what else do you know about the little orange square, it's all 90 degrees! Therefore, those are tools now to help you determine other measurement, that is one method; and the center point strategy is another....

These discussions begin to show that the lesson study experience made a positive difference for some of the teachers. Zsa's questioning about how much she should tell the students to support all learners and Jesse's response indicate that a collaborative culture in which to rethink their teaching was established by these teachers. Reteaching was planned for the next academic year, since the three teachers would be continuing to teach 5th grade.

Discussion of Results

In discussing the mathematics content, the teachers came to realize that they needed to know about the students' prior knowledge and how to deal with that knowledge or lack thereof. In each focus lesson, the teachers discussed a particular unit that brought up several questions that moved back and forth from the general mathematics curriculum to the specific mathematics in the *Investigations* unit. At the same time, the teachers sought coherence in the nine units taught at 5th grade. At first, the teachers viewed the units as disjointed but due to the collaboration and sharing during the lesson study process, the teachers found links between the units. Two of the teachers made connections emphasizing the importance of facilitating mathematical content found in the unit on fractions, decimals, and percents before they taught a probability unit, since the numerical value of any chance happening lies between zero and one.

As the teachers planned the lessons and explored how to focus their observations, they generated several strong points about inquiry-based instruction. For example, creation of a safe learning environment, the importance of listening to students' discourse, the need to anticipate questions, and the ideal grouping structures for exploring, sharing, and summarizing an activity either in a whole-class setting, or as collaborative small-group work, or between pairs of students. Jesse's use of humor was also noted as supporting student learning.

By the end of the second round of lesson study, observational data indicated that Zsa's instruction changed by her actively involving students in the lesson and using students thinking to guide her instruction. Her instruction, thus, began to involve more inquiry components and her teaching demonstrated an increased knowledge of content and students and knowledge of content and teaching. Zsa pointed out that she needed to know strategies to engage students in activities, as well as ways of probing students to explain their strategies without speaking out for them. During the planning session, Zsa's need was explored and the teachers agreed that the focus in the lesson study would be to watch and listen to what students might be saying to their partners and whether math language was used. Thus, Zsa felt that if she noted what the students said to each other and explored this in a whole-class discussion, her instruction would become more student-centered and less teacher-centered. Debriefing data show that while Zsa was able to capture this idea, she also criticized herself for not reviewing more mathematics content. When planning the geometry lesson, Zsa pointed out that "after high school that was it. I didn't ever want to deal with angles again." Zsa confessed that she may downplay key concepts, also due to her lack of knowledge about the specific concept. "I mean, and I try and then again I might be downplaying it. I tend to downplay stuff that I shouldn't be just because, not that I don't think it is important but that someone else will teach it or should have taught it."

Nora participated in the focus lesson planning and discussions but took very little of that planning into her own classroom. Her teaching strategies remained teacher-centered and unchanged during the time of the study even though she was using a standards-based curriculum. Nora told the students what to do and gave little opportunity for the students to construct knowledge through their own exploration, posing or asking questions, and group work. She held whole-class discussions in most lessons and remained the authority in these discussions. She did not use the games, choice time activities, or many of the tools in *Investigations*. There was no humor during mathematics time and the students spoke to the whole class only if she requested. She stated, "Once math starts, nobody's talking."

Table 2 Summary of teacher change

Teacher	Change	Evidence
Nora	No change	Had teacher-led instruction with minimal student-to-student interaction
Zsa	Implemented new instructional strategies	Involved students in learning; incorporated small groups of four. Listened and used students thinking in instruction
Jesse	Collaborative leader and coach	Posed many critical thinking questions during lesson study meetings, i.e., "what do you want us to look at?" He taught the first lesson study cycle. Had several mathematical charts and artifacts; displayed student's mathematical work on notice board

Jesse's practice matched the criteria for exemplary inquiry-based instruction as documented by the *Horizon Classroom Observation Protocol* (2003). He facilitated the lesson launch, sustained exploration and summarized the lesson, and created a safe environment with humor, which is conducive for discussion. He established a classroom culture of students' listening and responding to each another. He used co-operative small groups, games, explorations, high-level questions, and reflection as key points in his instruction before and after the lesson study. Jesse did not appear to improve his own teaching abilities through the lesson study experience. However, the collaboration gave Jesse an opportunity to take on a teacher leadership role. He was the first to volunteer to lead in the first focus lesson that helped create a collaborative environment for all the teachers. Zsa noted that she learned from his expertise not only as they planned and debriefed about the lessons but also from the observation of his classroom.

Conclusion

The teachers in the beginning of this study employed different teaching practices despite teaching in the same school and teaching from the same curriculum. Ball (1996) observed that however clear and illustrative curricula materials are, they still fail to "provide guidance on the specifics of day-to-day, minute-to-minute practice" (p. 502) or on what English (2002) refers to as "issues of significance to the classroom" (p. 7). Remillard and Byrans (2004) found that teachers' experiences impact how they use standards-based curricula and that teachers that collaborate with each other are apt to establish new classroom instructional approaches. The Blum et al. (2005) study adds that professional development programs that are teacher-led and immersed in actual classroom lessons are effective in supporting teacher change. Analysis of additional data collected by the first author of this study indicated that instruction in the three classrooms remained dissimilar despite their lesson study experience. However, two of the teachers did begin to grow professionally. Zsa began to develop an understanding of the student-centered approaches implied in the *Investigations* curriculum, and she was beginning to strengthen her understanding of the mathematics needed to teach each lesson. For Jesse the experience reaffirmed his instructional decisions in creating an effective learning environment for students. Jesse's growth came in his willingness to be seen both as a collaborator and as a leader among his colleagues.

An assumption by the authors of this study was that an inquiry-based instruction model as conveyed in *Investigations* materials would be a good fit for Nora, as she was the newly graduated participant. As a recent graduate from a preservice program, it was assumed that Nora would be more familiar with inquiry-based practice. This, however, was not the case. Nora participated in the focus lesson planning and discussions, but took very little of that planning into her own classroom. As seen, collaboration is necessary but not sufficient in fostering change. Nora taught day-to-day doing little what the materials required in instruction. She held to ideas from her prior experience. Besides the time constraints levied on her from her graduate school course work, Nora was the only temporary hire among the 5th-grade teachers. Her short-lived job situation may have led to a lack of commitment. She did what was required of her without getting fully involved with its details. The other experienced and tenured teachers tried to align their teaching practice with the school's mission; and in this alignment, they constantly checked to see what they had to negotiate between the school's, the parents', and their own perceptions of effective instruction. This study did find that the teachers' collaboration in team meetings and lesson study assisted in reducing some variations in classroom instruction among Zsa's inclusive classroom and Jesse's class to the benefit of the students in both classrooms. Both Zsa and Jesse noted that they would support further work in developing a lesson study experience in the coming year.

References

Ball, D. L. (1996). Teacher learning and the mathematics reforms: What we think we know and what we need to learn. *Phi Delta Kappan, 77*, 500–508.

Ball, D. L., & Bass, H. (2000). Interweaving content and pedagogy in teaching and learning to teach: Knowing and using mathematics. In J. Boaler (Ed.), *Multiple perspectives on mathematics teaching and learning* (pp. 83–104). Westport: Ablex Publishing.

Ball, D. L., Thames, M. H., & Phelps, G. (2008). Content knowledge for teaching: What makes it special? *Journal of Teacher Education, 59*, 389–407.

Blum, H. T., Yocom, D. J., & McLaughlin, M. (2005). Professional development: When teachers plan and deliver their own. *Rural Special Education Quarterly, 24*(2), 18–21.

Bogdan, R. C., & Biklen, S. K. (2003). *Qualitative research for education: An introduction to theory and methods* (4th ed.). Boston: Allyn and Bacon.

Cramer, K. A., Post, T. R., & del Mas, R. C. (2002). Initial fraction learning by fourth- and fifth-grade students: A comparison of the effects of using commercial curricula with the effects of using the Rational Number Project Curriculum. *Journal for Research in Mathematics Education, 33*, 111–144.

English, L. D. (2002). Priority themes and issues in international research in mathematics education. In L. D. English (Ed.), *Handbook of International Research in Mathematics Education* (pp. 3–15). Mahwah: Lawrence Erlbaum Associates.

Fernandez, C. (2005). Lesson study: A means for elementary teachers to develop the knowledge of mathematics needed for reform-minded teaching? *Mathematical Thinking and Learning, 7*(4), 265–289.

Fernandez, C., Cannon, J., & Chokshi, S. (2003). A US-Japan lesson study collaboration reveals critical lenses for examining practice. *Teaching and Teacher Education, 19*, 171–185.

Flower, J. (1998). *A study of proportional reasoning as it relates to the development of multiplication concepts*. Unpublished doctoral dissertation, University of Michigan, Ann Arbor.

Frid, S. (1994). Negotiation of meaning in mathematics classrooms: A study of two year five classes. *Proceedings of the 17th Annual Conference of the Mathematics Education Research Group of Australasia, 1*, 271–278.

Goodrow, A. M. (1998a). *Children's construction of number sense in traditional, constructivist, and mixed classrooms*. Unpublished doctoral dissertation, Tufts University, Medford.

Horizon Classroom Observation Protocol. (2003). http://www.horizon-research.com/instruments/clas/cop.pdf. Accessed 11 Sept 2009.

Kamina, P. A. (2005). *Teachers' perceptions and practices of Investigation curriculum: A case study of fifth grade in an urban school.* Unpublished Dissertation, Syracuse University, Syracuse.

Mokros, J. (2000). *The Investigations curriculum and children's understanding of whole number operations.* TERC Web site: http://www2.lab.brown.edu/investigations/resources/studresults/iccuwno.html. Accessed 11 Sept 2003.

Mokros, J. (2003). Learning to reason numerically: The impact of *Investigation.* In S. L. Senk & D. R. Thompson (Eds.), *Standards-based school mathematics curricula: What are they? What do students learn?* (pp. 109–131). Mahwah: Lawrence Erlbaum Associates.

National Council of Teachers of Mathematics. (2000). *Principles and standards for school mathematics*. Reston: Author.

Putnam, R. T. (2003). Commentary on four elementary mathematics curricula. In S. L. Senk & D. R. Thompson (Eds.), *Standards-based school mathematics curricula: What are they? What do students learn?* (pp. 161–178). Mahwah: Lawrence Erlbaum Associates.

Remillard, J. T., & Bryans, M. B. (2004). Teachers' orientations toward mathematics curriculum materials: Implications for teacher learning. *Journal for Research in Mathematics Education, 35*, 352–388.

Reys, R., Reys, B., Lapan, R., & Holliday, G. (2003). Assessing the impact of *standards*-based middle grades mathematics curriculum materials on student achievement. *Journal for Research in Mathematics Education,* 34, 74–95.

Shulman, L. S. (1987). Knowledge and teaching: Foundations of the new reform. *Harvard Educational Review, 57*, 1–22.

TERC [Technical Education Research Centers]. (1998). *Investigations in numbers, data, and space*. Menlo Park: Dale Seymour.

Response to Part IV: Seeing the Whole Iceberg—The Critical Role of Tasks, Inquiry Stance, and Teacher Learning in Lesson Study

Catherine Lewis

Taken together, the three chapters in this part help to illuminate several critical features of lesson study that have not been strongly emphasized in prior accounts of lesson study. Brian Doig, Susie Groves, and Toshiakira Fujii liken the process of task design and selection within lesson study to an "iceberg." They write:

> We argue that Japanese educators place a strong emphasis on task design and selection and that this effort is largely ignored by non-Japanese adapters of Lesson Study, possibly because the effort involved may be almost invisible, in the way that about 90% of an iceberg is invisible, with all of our attention going to its visible tip.

The authors further explain that teachers must attend carefully to the mathematics emphasized by a task, in order "to explicate all the mathematical concepts and understandings that make a particular task or problem 'float' mathematically."

The metaphor of the iceberg is useful for exploring not just task design and selection within lesson study, but also for exploring the features of lesson study discussed by the other two chapters in this part: inquiry stance, as discussed by Jacqueline Sack and Irma Vazquez; and teachers' learning dispositions, as discussed by Penina Kamina and Patricia Tinto. Each of these three components—a well-designed task, inquiry stance, and teacher learning dispositions—is integral to lesson study, as the three chapters reveal.

First, with respect to mathematical tasks, Doig, Groves, and Fujii provide three in-depth examples of tasks that can be used in lesson study, showing how they play out in actual classroom lessons. The authors identify four types of tasks used in Japanese lesson study, i.e., tasks that: directly address mathematical concepts, develop mathematical thinking processes, have been chosen based on rigorous examination of scope and sequence, and address known misconceptions. Have you ever experienced a lesson that directly builds student understanding of the *concept* of a circle? Have you considered what central ideas about a circle you would want grade-three elementary students to grasp, and what experiences would help students to build these ideas? If the answer to any of these questions is "no," you will be in for a treat

C. Lewis (✉)
Mills College, Oakland, CA, USA
e-mail: clewis@mills.edu

L. C. Hart et al. (eds.), *Lesson Study Research and Practice in Mathematics Education*, DOI 10.1007/978-90-481-9941-9_18, as you read how students investigate how to position children fairly in a game of quoits, and in the process build their basic understanding of a circle.

The tasks laid out in the chapter by Doig, Groves and Fujii remind me of research lessons I have seen in Japanese classrooms that I remember clearly more than a decade later, because they so changed my own mathematical thinking. For example, in one elementary grade-three lesson, Japanese students discovered *pi* by measuring and recording the circumference and diameter of circles, ranging in size from coins to hula hoops to a large circle marked on the playground. I will never forget the students' surprise as they recorded and shared their data and were astonished to find that the circumference was always "a little more than 3 times the diameter," even in circles that differed dramatically in size. This chapter leads us to imagine how different lesson study would be around the world if all mathematics teachers had access to such high-quality tasks as integral components of what the authors call "coherent sequences of lessons which are focused on the mathematics per se." As they go on to note, "While examples of curriculum materials with such an orientation exist, they are not widely seen outside the Netherlands…or Japan." While I imagine that some readers who have developed curriculum materials will find this point very hard to swallow, I think it is worth our wholehearted consideration. Many interesting *tasks* have been developed by researchers and educators in many countries, but interesting tasks do not necessarily add up to "coherent sequences of lessons which are focused on the mathematics per se." Too often, interesting tasks become side trips from the scope and sequence or enrichment activities, rather than the backbone of the curriculum.

The chapter by Doig, Groves, and Fujii notes that tasks are not "silver bullets" that automatically enable good lesson study or good instruction. Teachers must carefully study the tasks and their relationship to the scope and sequence of the curriculum, and must bring much pedagogical knowledge to the work of task implementation. As the authors conclude:

> …the strengths of lesson study rest on two significant bases. The first is the detailed planning of the lessons, which, in turn, is based on deep reflection on the mathematics and pedagogy. While many lesson study groups outside Japan focus attention on the mathematics, often this is at the expense of the pedagogy, or vice versa. It is critical that a balance be maintained.

The chapter by Kamina and Tinto offers insight into how teachers negotiate the mathematical and pedagogical demands of lesson study. Their case documents how a group of teachers used lesson study to support their implementation of a reform curriculum (*Investigations*) that challenged teachers' pedagogical beliefs and mathematical content knowledge. As the chapter documents, the three teachers in the lesson study group brought different backgrounds to the collaborative work, and they learned different things from the lesson study experience. Some learnings were mathematical, whereas others were pedagogical, and still others had to do with development of personal qualities such as leadership. As a Californian teacher once remarked to me, lesson study is differentiated learning for teachers. In the group studied by Kamina and Tinto, "the lesson study process facilitated the sharing of expertise from the experienced teacher to the less experienced teachers," with the lesson plans based on the *Investigations* curriculum providing shared objects around which both pedagogical and mathematical knowledge could be exchanged among teachers.

The chapter by Kamina and Tinto reminds us that lesson study is not "one size fits all" professional development in which we expect each teacher to change in identical ways. Their qualitative research attends to the different learning trajectory and outcomes of each participant, underlining the challenges that await researchers who aspire to conduct large-scale impact studies of lesson study. What was common across all the teachers in the group is that they showed the disposition to learn in some way—about mathematics content, pedagogy, or nurturing colleagues' practice—although not necessarily in every way that they *might* have.

Jacqueline Sack and Irma Vasquez focus their chapter on lesson study as a "method for developing and sustaining an inquiry stance for teachers." They point out the substantial overlap between lesson study and design research as inquiry processes that entail cycles of lesson design, observation, analysis, reflection, and redesign. In both lesson study and design research, careful attention to student thinking and work is a pivotal feature, providing insights into needed redesigns in tasks and pedagogy. Many lesson study groups (and probably many design researchers) set out with an expectation that they will produce, hone, and finalize lesson plans in a modest period of time. But as Sack and Vasquez point out, finalized lesson plans are an elusive goal:

> When the project began in 2007, the research team believed that they might have a set of publishable lesson plans by the end of the second or third year. Now in the middle of Year 3, they realize that their lesson plans continue to be refined and evolve according to the needs of each group of students. At this time, plans are still "under construction" and are only available to teachers who attend professional development sessions led by research team members. The team envisions going through additional iterations before they can begin to formalize the currently used lesson plans.

As all three chapters illustrate in different ways, high-quality lesson plans are not a "silver bullet" for improvement of instruction. Rather, they are one essential ingredient in the mix of ingredients that will support improvement of instruction. Taken together, the three chapters help us think about a theoretical model that captures lesson study's impact over time. While the graphic of lesson study (see Fig. 1 in the chapter by Murata, this volume) captures major features of the lesson study cycle, it does not capture what happens as lesson study shapes the curriculum, individual teachers, and teaching culture over time, ideas that Fig. 1 illustrates.

Figure 1 reminds us that lesson study does not operate simply by improving a lesson plan that then improves instruction. Lesson study should also support the growth of individual teachers and of the teaching community—for example, development of norms that expect inquiry, de-privatization of practice, and self-improvement effort. From the issues raised by these chapters and by Fig. 1, I would like to pose some challenges to those of us involved in lesson study and mathematics education research outside Japan.

First, how can we make high-quality mathematical tasks like those discussed by Doig, Groves, and Fujii central to lesson study and to the curriculum? As they note,

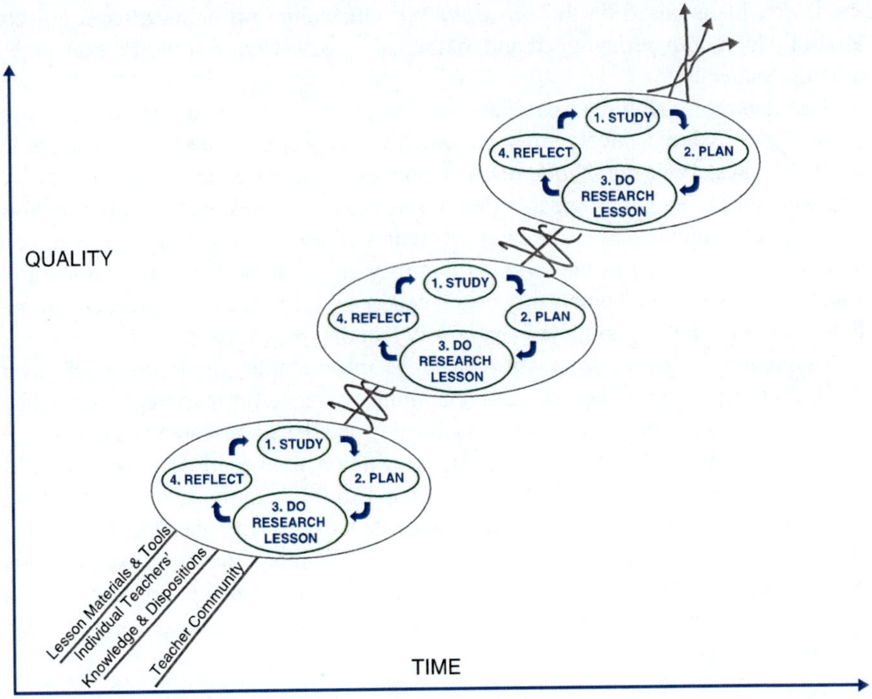

Fig. 1 Impact of lesson study on lesson materials, teachers, and teacher community over time

these tasks typically focus directly on key concepts, build mathematical processes, are based in rigorous examination of scope and sequence, and/or address misconceptions. In Japan, lesson study provides a route to examine and refine tasks, and over time the task examination and refinement conducted during lesson study influences textbooks and the scope and sequence of topics required in the national course of study (Lewis and Tsuchida 1997). So textbooks benefit over time from lesson study, incorporating increasingly high-quality tasks.

In the United States (the country with which I am most familiar), several barriers work against the refinement and broad use of high-quality mathematical tasks like those identified by Doig and colleagues. There is no widespread process like lesson study in which researchers and educators *together* examine mathematics tasks in actual classroom settings. So tasks may become well known because they are used in a research study, curriculum, or assessment, without public discussion of whether and how these tasks relate to central mathematical concepts and key elements of the scope and sequence. Were teaching time unlimited and curriculum coherence unimportant, it might be fine to have researchers focus their investigations on interesting tasks whose relationship to the scope and sequence has not been carefully examined. But too often, these new mathematical tasks remain peripheral to the curriculum or get added onto an already overloaded curriculum.

(I am not sure which is worse.) To take the example of fractions, there are vast resources on fractions available to US teachers, including many interesting tasks developed by researchers. But there is little consensus about the relationship between these tasks and the sequence of learning experiences that students should have, contributing to the problem of overloaded and incoherent curriculum. For example, while two major Japanese textbook series use the same four representations of fractions throughout the elementary years (linear measurement, area measurement, liquid measurement, and number line) and lay out a compelling argument for these particular choices, two US textbook series use a total of 25 representations (Lewis et al. 2010). Which approach do you think students will find more coherent?

Further, there is often an emphasis on originality and the "cool new task" when US researchers talk about mathematics tasks. Researchers are less likely to focus on analysis and refinement of the existing tasks in the curriculum than on invention of something new—perhaps unavoidably, because only the latter is likely to yield research publications. The demands of the research world (for individual creativity, for tasks that can be published as original work) may collide with the larger public need for analysis and refinement of the shared curriculum in ways that bring both teachers' and researchers' knowledge to bear upon it.

A second challenge for those of us conducting lesson study outside Japan is to represent lesson study in ways that better capture the processes of teacher learning, inquiry, and teacher community development that are at the core of lesson study (along with the development of lesson plans and tasks). The term "lesson study" connotes, to many non-Japanese, a focus on the lesson plan. While the lesson plan is certainly central to lesson study, the word "lesson" (jugyou、授業) in Japanese always refers to *live* lessons, not to something that can be captured on paper (as it sometimes does in English, when we hand someone a piece of paper and say "take a look at my lesson"). So the "lesson" in "lesson study" might better be thought of as "instruction," entailing pedagogy, teacher knowledge, and cultural norms about teaching and learning. While existing accounts of lesson study help readers understand how lesson plans are refined (Fernandez and Yoshida 2004), the equally important role of lesson study in refining teachers' knowledge and in building teacher community are less often noticed (Perry and Lewis 2008; Lewis et al. 2009). Together, these three chapters remind us that lesson study simultaneously rests on, and promotes, not just the development of lesson plans, but the development of the curriculum (instantiated in a sequence of mathematical tasks), the development of individual teachers' knowledge and dispositions, and the development of a teaching community whose norms and practices support continued growth of teachers and of the curriculum.

Acknowledgements This material is based upon work supported by the National Science Foundation under Grant No. 0207259 and Department of Education Institute for Education Sciences, Grant No. R308A960003. Any opinions, findings, and conclusions or recommendations expressed in this material are those of the author(s) and do not necessarily reflect the views of the National Science Foundation.

References

Fernandez, C., & Yoshida, M. (2004). *Lesson study: A case of a Japanese approach to improving instruction through school-based teacher development.* Mahwah: Lawrence Erlbaum.

Lewis, C., Perry, R. et al. (2009). Improving mathematics instruction through lesson study: A theoretical model and North American case. *Journal of Mathematics Teacher Education 12*(4), 285–304.

Lewis, C., Perry, R., Friedkin, S., & Baker, E. (2010, May). *Learning fractions in a linear measurement context: Development and fieldtests of a lesson study intervention.* Paper presented at the Annual Meeting of the American Educational Research Association, Denver, CO.

Lewis, C., & Tsuchida, I. (1997). Planned educational change in Japan: The case of elementary science instruction. *Journal of Educational Policy 12*(5), 313–331.

Perry, R., & Lewis, C. (2008). What is successful adaptation of lesson study in the U.S.? *Journal of Educational Change 10*(4), 365–391.

Part V
Ideas for Developing Mathematical Pedagogical Content Knowledge Through Lesson Study

What's Going on Backstage? Revealing the Work of Lesson Study with Mathematics Teachers

Catherine D. Bruce and Mary S. Ladky

In their seminal 2006 lesson study article, Lewis et al. (2006) highlighted the four key stages of the Japanese lesson study cycle (goal setting, planning, lesson implementation, and reflection) and using excerpts culled from field notes and transcripts provided evidence of teacher activity at each stage in the cycle. Researchers, such as those in the Psychology of Mathematics Educators-North America (PME-NA) Lesson Study Working Group, and participating teachers have found this model to be critically important in guiding the lesson study process. As lesson study moves into its adolescence in North America, our understanding of the subtleties of each stage is increasing. In this chapter, we explore some of the less documented activity, what we are calling the "backstage work," which was identified by mathematics teachers in our two-year study (see also Ross & Bruce, 2008) as they moved between and from stage to stage. Twelve teachers, who ranged in experience levels from novice to over 20 years, engaged in two lesson study cycles. Their focus was on the use of interactive whiteboards and manipulatives as supports to student learning in mathematics classes. Three of the school sites were secondary schools (one rural and two urban) and the fourth site was an urban elementary school. All schools had predominantly English-first-language students. As researchers, our goal was to locate and better understand the value of the informal professional learning activities teachers engage in while participating in a formal lesson study.

During focus group interviews, which occurred on three occasions between September 2007 and June 2008, researchers asked the 12 mathematics teachers to describe the informal activities that took place between the formal stages of the cycle. Further, in video-recorded and transcribed whole-group discussions (Lerman 2001), teachers built a parallel model describing the informal activities they engaged in between the formal stages of the lesson study (see Fig. 1). This chapter highlights what goes on backstage, thereby contributing to an enhanced, more nuanced understanding of the lesson study cycle. Participant voices featured in this chapter are based on transcripts of the interviews and meetings, and illustrate the backstage work from the perspective of those deeply engaged in the lesson study process.

C. D. Bruce (✉)
Trent University, Peterborough, Canada
e-mail: cathybruce@trentu.ca

L. C. Hart et al. (eds.), *Lesson Study Research and Practice in Mathematics Education*,
DOI 10.1007/978-90-481-9941-9_19, © Springer Science+Business Media B.V. 2011

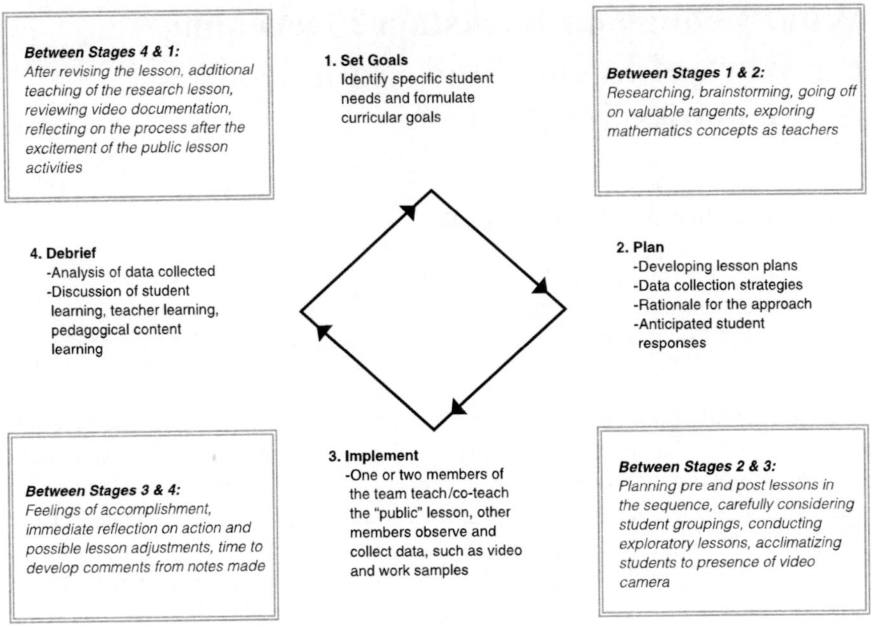

Fig. 1 Revealing the backstage work of the lesson study cycle

In Preparation for Stage One

In the first formal step of the Japanese lesson study cycle, teachers work together to establish curricular goals for their mathematics lesson planning. This goal often stems from data that teachers previously gathered. For example, the elementary teacher team in our study used PRIME (2005) diagnostic materials to assess which curricular areas in mathematics students were struggling with the most. In this instance, the area identified was understanding growing linear patterns. These teachers were interested in finding ways to structure tasks using mathematical models, manipulatives, and interactive whiteboards to illustrate and manipulate growing linear patterns. One of the secondary teacher teams decided to select a curricular area that their students had struggled with from year to year and that they determined could be more effectively introduced with attention to lesson design. The problem identified related to solving linear systems. Again this teacher group wanted to physically represent x and y values (including unknown values) and finds ways for students to manipulate these physical materials to solve equations. The interactive whiteboard became a method for students to share their solutions using dynamic software. Between the first goal-setting stage and the second stage of planning a lesson based on curricular goals, many backstage activities occurred.

Between Stages 1 and 2

These mathematics teachers identified four main backstage activities between stage 1 (goal setting) and stage 2 (planning) of the lesson study cycle. They are: (1) "searching," which included researching using the internet, data-bases, and teacher resources about the topic in focus; (2) "conceptualizing," which included brainstorming, self-talk, and informal conversations, as well as exploring valuable tangents; (3) "investigating," which included exploring the use of manipulatives and technological tools with students to expand the teachers' and students' repertoire; and (4) "monitoring," which included keeping up with details such as on-going student assessment which provided insights into student learning and assisted in the planning of lessons.

Between stages 1 and 2, teachers frequently emailed one another to clarify thinking and to question one another.

> I guess we could phrase it as focused homework, in the sense that we would meet together, get so far and say okay, where do we need to be before we meet again? So you do that [email], you do that [make notes], you do that [find a resource], and we'd go away and bring the pieces back together.

Further, teachers began establishing and developing routines with students preparing them for the research lesson. This preparation also included familiarizing students with the use of the interactive whiteboard and introducing the use of the video camera in the classroom.

> We had to spend time developing some routines. When they come into the classroom for this activity, what routine are they going into to make sure that everything flows well? Picking up the laptops, putting them away, warm up....

Teachers also prepared by determining how to use manipulatives for maximum student learning given their lesson goals:

> We also spent time, and actually, once we made the decision about solving the equation lesson, how were we going to use concrete manipulatives and what concrete manipulatives we were going to use? And then we had to play to figure it out.

The teachers explained that the time they were provided for "playing" and "figuring out" was deeply connected to their own understanding of mathematical concepts and how those concepts could be represented and investigated by students: "When we were at Trent [University], that whole session on math was just, was just ourselves understanding...getting into the math." Adequate time to develop conceptual understanding is a critical component for teachers whose own mathematical knowledge may not be deeply developed.

Between Stages 2 and 3

The teachers identified backstage activities that continued on after stage 2 (planning) and before stage 3 (lesson implementation). These included emailing one another with questions and suggestions (one team used Wiki space for this, for example), further experimenting with manipulatives and interactive whiteboard use, and continued acclimatization of students to video camera use in the classroom.

In addition to these on-going activities, the teachers also described new activities which required significant amounts of time dedicated to developing lessons that would precede and follow the research lesson. Participants spent many sessions together planning related lessons leading up to and following the research lesson which, in essence, became a sequence of lessons over several mathematics classes. Teachers agreed that the planning of the lesson sequences felt painfully slow but that the process was absolutely necessary:

> We did a three-lesson sequence, so…we had to plan for those pre-lessons too, [and] actually, we did a lot of planning on those. So that's a huge use of time. But you need it—we really needed that.

The teachers were committed to documenting the full lesson sequence because one of their primary goals was to provide students with multiple opportunities to learn and understand complex mathematical ideas. Therefore, the entire lesson sequence that teachers developed was included in the lesson study packages for distribution on the day of the public lesson. This gave participants significant information about what students experienced prior to the research lesson. Also of note is how teachers described their deliberations over optimal student groupings for the specific tasks being planned: "We didn't just throw the kids together in groups. We sat and decided which students we should put together to optimize the outcome."

Teachers also engaged in exploratory lessons where they experimented with teaching strategies to see if they were effective for students. Exploratory lessons allowed the teachers to try out individual components of the research lesson and specific teaching strategies they were unsure about. During participant discussions about the exploratory lesson activity of their team, the following recorded discussion took place which illustrates the value of these exploratory lessons:

Teacher 1:	We've been fine tuning as we've gone through. You know we started with a 7/8 [grade] class, then we went to the 7 [grade] class, we made some changes between those two, and then um…we jumped to the 9 [grade] academic, and then we made some changes again to suit the 9 [grade] applied students. Some great things happened in the Grade 9 academic that came out, so we incorporated that into this next lesson…we've evolved.
Teacher 2:	And we changed the "Minds On". We tried to improve student understanding for the definition of volume, between the grade 7/8 and grade 7, so that was when we started to really make that change.
Researcher:	The first "Minds On" was the one you had the circumference…
Teacher 3:	So now it's focused upon an aquarium with a fish…
Researcher:	Prism displacement…
Teacher 4:	Much better.
Teacher 5:	We even fine tuned our grouping strategies too. We realized with a group of three, one student was always…it was easier to pull away from the activity, and groups of four we found the two weak students would go together and two strong students would go together, so with the group of four we had variants of two, now in Susan's class we'll just have pairs.

Between Stages 3 and 4

The public lesson (stage 3) days were high energy, involving all teachers from the lesson study planning activities as well as invited guests and school administration. After the public lesson, teachers agreed that together they experienced collective accomplishment: a sense of "we did it!" All participating teachers commented on how important it was for each of them to share this sense of accomplishment with their team members directly after the public lesson and before the formal debrief in stage 4 of the cycle.

> And there was... a sigh of relief that kind of implies, thank god it's over, but also and I think more importantly, it was thank god I did that! There are people who will teach forever and never have an opportunity to be a part of that. Thank goodness I was a part of that.

This sense of accomplishment was inextricably coupled with teacher thoughts about the "I should-haves and I wish I hads and the oops..." which are key instances illustrating reflection-on-action (Schoen 1983). As Garcia et al. (2006) point out:

> Mathematics teachers' reflection on teaching situations is an important process providing information that contributes to our understanding of their professional knowledge. This kind of reflection can be generated in many different ways: (i) spontaneously with the help of researchers (Jaworski, 1998), (ii) included in mathematics teacher education programmes and professional development through the use of narratives (Ponte et al., 1997; Schifter, 1996) or (iii) in research projects where teachers and researchers collaborate. (Garcıa and Llinares, 1999; Llinares, 1999). (p. 2)

Directly following the public lesson, the observers were given 15 minutes to reflect and organize their notes. Each observer (including all members of the teacher planning team) made summary notes based on their observations so that they had a coherent set of comments to make during the debriefing. The teachers explained that they appreciated this time to "gather their thoughts quietly" before launching into the debriefing. These reflections-on-action were then shared using specific protocols in the debriefing (stage 4).

The 15-minute reflection-on-action that participating teachers engaged in before the formal debriefing included recording summaries about specific student observations, noting logistics and effectiveness of the selected of mathematics materials, summarizing tallies of the recorded questions asked by students and teachers by type, and organizing data collected about gender participation in the lesson. The lead teacher for the lesson also reflected on and recorded notes about surprises she or he encountered during the lesson and these were subsequently shared as the first component of the debriefing stage. The following example highlights one of four surprises noted by the lead teacher in a lesson on linear functions using a problem about trapezoid tables and chairs:

> It took some conscious decisions throughout that and, for example, just to kind of highlight one thing that happened: When I gave them the manipulatives, I was surprised that not very many people knew what I was asking, and I was like, "Oh, obviously I wasn't clear enough", because they were—some were just using the tiles, they didn't even look at the trapezoids, and I thought, "It's a trapezoid—we just talked about trapezoid tables—obviously that's not clear". So I kind of had to go do some little interventions.

Between Stages 4 and 1

After the formal debriefing session (stage 4) where the teachers discussed the lesson and observers made comments about the lesson, the planning teachers regrouped informally as a small team. One teacher expressed how valuable it was to have these opportunities for "small group talk" during which the teachers agreed to designing follow-up lessons, which were found to be "invaluable."

The teachers immediately began planning the follow-up lessons based on the observations of the public lesson. This included revising plans that the teachers had already developed and teaching the research lesson to any students who had not yet had the opportunity to participate in the lesson. In some cases, the teachers asked students who had participated in the public lesson to teach their same-grade peers who were not part of the public lesson activity. A number of participants described these situations as opportunities for students to consolidate and demonstrate their understanding:

> The interactive whiteboard screen came up, they remembered what we'd done [as] it had just been the day before or two days earlier and they taught it to their peers. That was interesting. A lot of them soaked it up.

Finally, teachers described how they reflected on the public lesson more deeply over time. For example, in one case, the teacher who delivered the public lesson sat down with the researchers to watch the videotaped lesson (a technique encouraged by Lerman 2001). The researchers asked the teacher to "think aloud" as she was watching the video in order to audio-record the teacher's comments. After viewing the lesson, researchers asked the teacher what value there was in watching the videotaped research lesson. The teacher replied as follows:

> I feel like you don't really get a huge chance to digest everything in that time afterward [during the debrief]…you're on a timeline too…you have to all get back to class and everything. I guess this [watching the video] offered another chance to talk about it. And you [researcher] pulling questions out the way that you did really made me think about it differently and probably better than I would have on my own, more deeply. So it almost added to that sense of the debrief. It gives you a little time to digest when you really see it, I mean, that's pretty neat. Because you're not watching everything that's going on when you're teaching right? So it's adding on to that. And seeing the kids too, being able to do that, step back, and watching them—remembering what the lesson was about, and definitely talking about that—but watching them in action too, and just having that different perspective.

In addition to these overall positive outcomes, researchers and participants in this study recognized the backstage activities as essential building blocks of a successful lesson study cycle. These more informal interactions and processes represent significant work and commitment of the teacher teams in supporting one another. The participating teachers frequently cited their incidental planning meetings as being an important feature in the successful deployment of lesson study cycle. A majority of these meetings occurred casually over lunches, during recesses, and after school hours (in addition to formal facilitated meetings with release time). We believe that more research is required to further our understanding of how these backstage activities relate to the nuanced dynamics of a successful lesson study team as described in the chapter by Meyer and Wilkerson, particularly in relation to mathematics learning. For example, do teams who work together well, both with one another but also with their administrators and researchers, engage more frequently in backstage activities than teams who have greater difficulty in achieving their lesson study goals? Do successful teams have more opportunities to encounter one another on a more regular basis in order to communicate informally? And if so, what can researchers and administrators do to help facilitate these important backstage activities among school teams? Acknowledging that these backstage activities occur and that they are valuable is a crucial first step. We believe that making these backstage activities more explicit for those undertaking lesson study may contribute to a deeper understanding of the types of work required for high quality learning for all participants in lesson study.

References

Garcia, M., Sanchez, V., & Escudero, I. (2006). Learning through reflection in mathematics teacher education. *Educational Studies in Mathematics, 64*, 1–17.

Lerman, S. (2001). A review of research perspectives on mathematics teacher education. In F. L. Lin & T. Cooney (Eds.), *Making sense of mathematics teacher education* (pp. 33–52). Dordrecht: Kluwer.

Lewis, C., Perry, R., & Murata, A. (2006). How should research contribute to instructional improvement? The case of lesson study. *Educational Researcher, 35*(3), 3–14.

PRIME (Professional Resources and Instruction for Mathematics Educators). (2005). *Number and operations diagnostic tests*. Toronto: Thomson Nelson.

Ross, J. A., & Bruce, C. (2008). *Quantitative report on lesson study GAINS initiative*. Report to the Ministry of Education, Ontario. (To access report, contact: Dr. John A. Ross, Professor & Field Centre Head, OISE/UT Trent Valley Centre, PO Box 7190, 1994 Fisher Dr., Peterborough, ON K9J 7A1).

Schoen, D. (1983). *The reflective practitioner*. London: Temple Smith.

Learning from Lesson Study: Power Distribution in a Community of Practice

Dolores Corcoran

Introduction

When lesson study in relation to mathematics teaching is discussed in English, it most often emphasizes the potential for development of mathematical thinking by means of innovative and carefully orchestrated lessons (Sugawara 2008). Apart from a number of teacher development studies (see Fernandez et al. 2003; Fernández 2005), much of the literature is directed at "how to" conduct lesson study as opposed to describing how/why lesson study works. In 2006, Lewis, Perry and Murata called for research into the issue (Lewis et al. 2006). They formulated two conjectures as to how lesson study works to bring about the improvement of teaching. Conjecture 1 suggests, "Lesson study improves instruction through the refinement of lesson plans." Conjecture 2 suggests, "Lesson study strengthens three pathways to instructional improvement: teachers' knowledge, teachers' commitment and community, and learning resources" (ibid, p. 5). My interpretation of lesson study favoured conjecture 2.

The Study

This chapter draws on data from a lesson study elective course in which six third-year Bachelor of Education students participated over two semesters. All were female; four were 'mature' students in that they had engaged with other career paths before embarking on a teacher education program. None had chosen to study mathematics since leaving secondary school, where two had studied higher-level mathematics. The course arose as the third tier of a three-tier research project which set out to investigate three questions: first, the mathematical knowledge which Irish student teachers bring to teacher education; second, how this impacts on their teaching of

D. Corcoran (✉)
St Patrick's College, Dublin City University, Dublin, Ireland
e-mail: dolores.corcoran@spd.dcu.ie

L. C. Hart et al. (eds.), *Lesson Study Research and Practice in Mathematics Education*,
DOI 10.1007/978-90-481-9941-9_20,

mathematics during the second-year teaching placement; and third, how mathematical knowledge for teaching might be developed more successfully (Corcoran 2008).

The Participants

The primary site of this research is an Irish college of education with an annual intake of 400 students who undertake a three-year Bachelor of Education (BEd) degree course. The course is popular, and entry is by competitive examination. As a result, all BEd students are high achievers academically, but not necessarily in mathematics, where entry requirements are quite low. There is no syllabus to mandate what mathematics pre-service primary teachers should learn. All students participate in two 'curriculum' mathematics methods courses, one twenty-four-hour module in second year and another in third year. Assessment of these courses is included with many others in the Education element of the BEd programme, where mathematics education can appear unimportant.

Of a possible 400 respondents, 125 opted to take a mathematics audit to indicate their mathematical knowledge for teaching. As a result of tier one of my study, it could be claimed that some of the student teacher participants exhibited 'secure' mathematical knowledge for teaching and some were 'insecure'. However, most students in the year cohort (N = 400) were not prepared to address the matter at all. These prospective teachers may be thought to have considered mathematics as "just one of the subjects I have to teach" (Brown and McNamara 2005). One such student was Treasa, whose participation is described later. She opted not to take the mathematics audit in tier one, did not participate in tier two, yet as a third-year student leader made a deliberate choice to engage with her feelings of uncertainty and discomfort with mathematics teaching by joining the lesson study elective course. Five of the six members of the lesson study group had chosen to take the mathematics audit with varying results. Róisín and Nóirín performed well in tier one, while audit scores for Bríd, Ethna and Finola were towards the lower end of the spectrum. All students in the cohort—including those who abstained from the mathematics audit—were expecting to teach primary mathematics during their next two assessed teaching placements and in their future teaching careers. Reluctance to audit their mathematics knowledge for teaching raises questions about the status of mathematics in their teaching repertoire.

For tier two of the study, twelve volunteers (including seven mathematically 'secure' students) were video recorded, each teaching a mathematics lesson during their second-year teaching placement. Bríd, Róisín and Ethna were participants in tier two. The Knowledge Quartet (KQ) (Rowland et al. 2005) was used as a theoretical lens through which to look at mathematical content knowledge at work in classroom teaching. The KQ is a framework of eighteen contributory codes that can be used to identify mathematical knowledge in teaching arranged along four dimensions—Foundation, Transformation, Connection and Contingency (ibid, pp. 265–266). Findings from this tier indicated that student teachers were working hard to teach well. However, individualistic planning approaches and disjointed mathematical thinking even among students with secure mathematical profiles seemed to result in hybridised forms of mathematics teaching, which may not facilitate pupils' learning of mathematics. The realisation that student teachers who identify themselves as 'good at maths' do not necessarily teach mathematics any better than those who identify themselves as 'not good at maths' led to the lesson study elective course outlined here.

The elective course was designed as an innovative means of developing mathematical knowledge for primary teaching using the lesson study protocol, with six research lessons video recorded and analysed over the academic year 2006–2007. Learning mathematics for teaching took place for community members through participation in on-going *negotiation of a shared enterprise* (Wenger 1998). There were three distinct aspects to the mathematics lesson study elective course. First, students engaged in 'doing' lesson study by participating with others in preparing, teaching and reflecting on mathematics lessons. Second, participants also engaged in 'doing' mathematics together regularly. Third, by 'being' lesson study elective students they participated in activities related to the elective course but not essential to lesson study, e.g. writing reflective journals, writing course assignments, viewing lesson study DVDs (Mills College Lesson Study Group 2000). The mandatory graded teaching placement for these students would take place for four weeks in the following spring term, while the lesson study elective offered student participants an opportunity to engage with research lessons in real classrooms without the pressure of being 'examined' as they did so. In this chapter, I draw on data from the three above-mentioned aspects of learning from practice and highlight their contribution to development of transformative power relations within the community. Each lesson study cycle revolved around two research lessons taught concurrently, where the whole group engaged in the preparation of and reflection on each lesson together. On research lesson days, the community divided into two groups and each 'teacher' was accompanied by at least two observers. Mathematical topics addressed included weight followed by fractions and division. See Table 1 for list of research lessons and Table 2 for descriptive synopses of lessons taught. There were three iterations of the lesson study cycle extending from September into the following spring. Data for the study include the video records of six 'research' mathematics lessons, reflective journals written by the students after each session, audio

Table 1 Research lessons taught during each lesson study cycle

LS cycle	School	Class/ages (years)	Topic	Student**
Cycle 1 OCT	St Peter's	4th/9–10	Weight	Treasa
	St Paul's*	4th/9–10	Weight	Finola
Cycle 2 NOV	St Peter's*	5th/10–11	Fractions	Bríd
	St Paul's	3rd/8–9	Fractions	Ethna
Cycle 3 DEC	St Peter's	3rd/8–9	Division	Róisín
	St Paul's*	5th/10–11	Fractions	Nóirín

* Researcher present

** Pseudonyms were used

Table 2 Descriptive synopses of six research lessons

St Peter's National School	St Paul's National School
Research lessons in Cycle One	
Grade 4	Grade 4
Treasa introduced a problem: "Would you be able to take your school bag on a Ryanair flight where there is a restriction of ten kilogrammes on carry-on luggage?" Children were asked to compare and order their school bags and record their findings on a scale, which Treasa drew on the board. Then children were introduced to a kilogramme of sugar as a reference weight and asked to select items from their schoolbag which they estimated might be the same weight as a kilogramme. Treasa introduced some kitchen scales and bathroom scales for children's use to check their estimates as they worked in groups. The lesson concluded with a whole class discussion of findings	Finola brought a number of food items to be weighed. The class teacher had already assembled a number of weighing instruments for the class to use. Finola stated that a child's schoolbag should not be more than a tenth of a child's body weight and posed questions about how these children's school bags measured up in this regard. She also established the kilogramme as a reference unit of weight and invited children to compare the weight of certain schoolbooks in relation to a kilogramme benchmark and to measure food items on various scales. Finola drew a scale, starting at 0 kg, on the board and invited children to tell her where to place their school bags according to their weight
Research lessons in Cycle Two	
Grade 5	Grade 3
Bríd set the scene by asking them to imagine it was one boy's birthday and that he had invited seven friends to share six pizzas with him. The lesson consisted of three parts. Children were given a teacher-made handout showing six identical circles on an A4 page, which they were invited to think of as pizzas. After some time working in pairs, the class was called to attend while some pairs were invited to the board to explain what they had done. Next, the children addressed a second similar but arguably more difficult problem, involving six pizzas divided between ten people, and later other pairs of children were invited to the board to explain their work. A plenary session concluded the lesson where the teacher talked the children through the process of adding a half and a quarter, and finally elicited why the size of the unit was an important element in dealing with fractions	Ethna had eleven nine-year-olds in her classroom, seated in two groups of four and one of three. She used a lesson already published by Marilyn Burns (1987) on the division of cookies among four people. She brought paper cut outs of cookies and had prepared worksheets for the children and enlarged versions to hang on the board as she explained the tasks. Children were to divide four, then five, three and seven cookies among four children. The lesson concluded with Ethna simply telling children the meaning of the symbols 1/2 and 1/4. She drew circles to represent 'wholes' on the board and shaded in the appropriate fractional pieces, finally asking the children to copy these representations into their copybooks

Table 2 (continued)

St Peter's National School	St Paul's National School
Research lessons in Cycle Three	
Grade 3	Grade 5
There were 24 children present in Róisín's third-grade classroom as she asked them to imagine organising a class outing. She divided the board into four columns and wrote 24 in a box in the centre at the top to represent the number of children present. Children were given worksheets she had designed. Children were asked to divide the 24 equally between three adult 'minders', represented by stick figures. Then they were asked to divide the class of 24 among four minders. This was the partition model for division. Róisín asked children to describe a couple of ways of doing each problem. Then the class was to be transported in cars to an amusement park. How many cars would be needed if there were four children per car? Suppose there were only three children allowed to travel in a car, how many cars would be needed? Here Róisín was introducing the quotative model for division. In conclusion, Róisín counted herself with the class to make up twenty-five people, and after selecting four other leaders, invited the children to arrange themselves in teams of five in order to go bowling. They were then asked to record and check their findings	Nóirín had a small number of fifth-grade children and she asked them to engage in problem solving. They were to work in pairs and Nóirín used the same scenario as Bríd's pizza party, but was less specific about whether the six pizzas were to be divided among seven or eight people. She distributed "ways of thinking sheets" (Lesh and Clarke 2000) but did not offer any representations to the children. After ten minutes, she invited two different pairs of children to the board to represent their solutions. The second problem involving six pizzas divided among ten people was posed and solved with greater success by the children. Nóirín followed the same routine of calling a few pairs to the board to explain their work, and drew the lesson to a close by introducing two large coloured circles, one divided in two halves with a line through the centre, the second divided into tenths. She attached these to the board so that the line drawn through the centre of each circle was vertical. Nóirín then 'taught' the class that a half was equivalent to five tenths by shading in the appropriate pieces in each circle and further demonstrated that 1/2 + 1/10 = 6/10 using the same representation

recordings of the planning and reflective meetings in cycle one, video recordings of the planning and reflective meetings in cycle two and audio recordings of the post-lesson reflection meetings in cycle three.

The student teachers' lesson plans and examples of children's work were also part of the data as were my observation notes for the three research lessons I observed first hand. The two school sites were chosen because the two principals were prepared to facilitate the research by accepting the student teachers as 'guest teachers', securing parental permission for lessons to be video recorded and inviting interested staff members to facilitate the research lessons.

The lesson study group of six student teachers themselves chose the mathematics topics and class levels they wanted to address, and in each cycle the team devised problem-based mathematics lessons in the light of their collective interpretation of content objectives from the *Primary Mathematics Curriculum* (Government of Ireland 1999a). On research lesson days they were welcomed into classrooms where the lesson study process was seen as an interesting novelty.

My Learning from Belonging to the Lesson Study Community

My field notes from the first three sessions associated with the first lesson study cycle indicate that I was conscious of playing different roles at different times in the process. I approached the elective course as course tutor acting primarily from a research perspective. I was anxious that the students would understand that we were co-researchers and distributed copies of a Wagner (1997) article about the implications of collaborative research, which was to be studied and discussed later. I also distributed copies of the Lewis and Tsuchida (1998) article to aid discussion of the potential of lesson study, and a single page outlining the Knowledge Quartet and its contributory codes as a framework for lesson analysis. While any or all of the KQ contributory codes could be used to identify teachers' mathematical knowledge that can be observed in the act of teaching and its impact on a particular lesson, I was suggesting their use to the group as a shared language for discussing teaching. Thus the KQ framework was intended to build community by becoming part of the "shared repertoire of ways of doing things" (ibid, pp. 82–84). In my capacity as researcher, I was also prepared to play Knowledgeable Other to the group by conducting reflective discussion about lessons taught. I relied largely on my identity as a long-time teacher of primary mathematics prepared to support potential teachers in their planning for teaching. Initially, I attempted to minimise or play down my identity as course tutor and stressed that I was acting as facilitator to help students organize themselves into a lesson study group. Respecting the power of the group, I was at first prepared to 'let them at it'. Exactly how to pitch the degree of my input became a dilemma for me, and while reflecting on the first lesson study cycle planning meeting, my mathematics education lecturer persona took over.

Lesson Study Cycle One

It was not surprising to find that prospective teachers chose the topic of weight for their first research lessons arising out of their having watched the *Can You Lift 100 kg?* DVD (Mills College Lesson Study Group 2000) at the previous session. During planning sessions one and two, I became very aware of the participants' difficulty in aligning content objectives from the curriculum with suitable contexts in which to base problems and activities designed to promote children's reasoning about the mathematical ideas underpinning the chosen objective. Treasa, for example, suggested a word problem, indicative of reliance on a traditional context for teaching the topic.

Bríd: How would we weigh one biscuit in a packet of biscuits? Is that what you mean?
Treasa: No, like five boxes of biscuits together weigh twelve kilograms, find the weight of each box. (Lesson study cycle 1_Preparation session 1)

Treasa and Finola—neither of whom had been in tier two of my study—volunteered to teach the first research lessons. Apropos of mathematics in our environment, I mentioned having noticed a mother walking her 10–11-year-old son to school and carrying his school bag. Nobody in the group picked up on this notion, and a more obvious introductory problem was chosen. After the first planning meeting I sent an email to Treasa and Finola outlining further suggestions for a realistic context for the weight lesson they had devised. Treasa resisted this and proceeded with the investigative lesson planned by the group: "Could you bring your school bag on a Ryanair flight where a ten kilograms weight restriction applies?" Finola adapted her lesson in line with my 'new' suggestion: "Children ought not to carry school bags that weigh more than a tenth of their body-weight. How does your schoolbag measure up?" Studying a transcript of the first planning session, I was struck by the distrust participants appeared to have in the efficacy of a problem-based mathematics lesson. Thinking that perhaps my course tutor input was more important than I had anticipated, I resolved to foreground my mathematics educator role more in the next planning session and came prepared with sample lesson plans, downloaded from a UK website (National Numeracy Strategy 2006). Later, however, I realized that I was not needed to be a 'teacher' to the group, although they continued to use my expertise and my access to mathematics education research as a resource. As a result, negotiating the boundaries of my role became an unexpected research theme that developed over time. By the third cycle of lesson study students were prepared to operate as a community of practice without my constant presence, and they noted questions that they would pose to a Knowledgeable Other at a later date.

Bríd, whose engagement with the collaborative nature of the lesson study process deepened when she opted to teach one of the research lessons in cycle two, commented on a shift in agency:

> For the second lesson [in cycle three] we focused on introducing Percentages to 5th class. However, it soon became clear that we were not clear how to introduce it. It is more complicated than I first thought and it was really only through lesson study that I came to realise it. (Journal 7_Bríd)

Alternative ways of approaching the topic of percentages were explored, and since in Bríd's opinion "we got ourselves into a mess", a group decision was taken to choose another lesson topic on that occasion, but with the intention "to look at it again with Dolores for guidance." In this cycle, the participants were forced to abandon teaching a lesson on percentages to eleven-year-olds, after spending a considerable time preparing it, because Nóirín balked at teaching the topic.

> At this stage I was becoming increasingly confused about the whole idea of percentages and I really couldn't see where we might be heading with this planning. Also, I was thinking at this stage about the fact that the children had very limited knowledge about percentages and I was worried that I was going to have to introduce them to the idea. I was anxious that I would only confuse them with it if I wasn't fully confident about it myself and that would not be a good way to introduce the children to a new topic. (Journal 7_Nóirín)

I infer from this that the prospective mathematics teacher identities had begun (a) to acknowledge limitations in knowing as remediable, (b) to resolve to seek appropriate help for further learning and (c) in the interim, to resource from within the community of practice alternative approaches to good mathematics teaching.

Relationships with Mathematics

While interpersonal relationships were important to everyone from the outset, it appeared that the student participants were more concerned about their own relationships with mathematics. These students accepted that posing realistic problems and focusing on children's responses were aspects of good mathematics teaching, yet because of their own fragile relationships with mathematics, these actions were challenging for them. At the outset, Bríd acknowledged her personal insecurities in relation to solving the simple problems we engaged with as a group at our first meeting.

> Personally, I find it very difficult to start the problem and always doubt my ability to do… we have to encourage children to use the way that's best for them, however, it scares me too—how will I know if the child is learning the basic content if I don't understand their approach? I feel uneasy about dealing with the wide scope of approaches and/or elements that children will use. (Journal 1_Bríd)

Bríd was at that stage in the third year of her teacher preparation course, having already qualified and practised as a teacher of two subjects—though not mathematics—at second level. I saw my task become one of making the lesson study elective sessions a safe place to question one's own and each other's mathematical ideas. The role of the tutor as Knowledgeable Other or "old-timer" (Lave and Wenger 1991, p. 57) developed in the selection and presentation of interesting mathematics to engage the group in the 'doing' mathematics aspect of the collective enterprise, and in drawing pedagogical inferences from events in the group setting. To facilitate the expression of mathematical thinking, I also sought to establish facilitative communication norms within the community of practice. These were not specifically spelled out. Rather, we commented on qualities of the environment—mutual engagement, trust, acceptance, openness to questioning, uncertainty and diversity—which when present, enhanced people's capacity to think mathematically. When working in community, all members had responsibility to the enterprise, but knowing when to deviate from the agenda or when to allow the discourse to continue uninterrupted presented an occasional dilemma for me. Working with the KQ as an analytic tool, my awareness of Contingency moments where I could have chosen to respond to someone's ideas or used unexpected opportunities grew as I listened to tape recordings of the first lesson study cycle sessions. I was struck by the similarity between the conversations among participants in the lesson study group and mathematical discourse in a lively classroom. Different approaches to mathematics were celebrated and discussed within the group, and tended to mask the fact that some prospective teachers were quicker and surer in proffering solutions than others. Some students commented in their journals on these differences, but our focus in the community was primarily on improving pedagogy. We sought constantly to make connections between intentions and actions of the teacher as well as between imagined and actual responses of the children, so the aspect of difference in mathematical facility within the group was minimised. Nonetheless, challenges in the communication of ideas also emerged as an important aspect of the lesson study enterprise. By becoming aware of, and seeking clarification of the many possible interpretations of a particular idea put forward by one member of the group others became more open to listening and learning.

Acknowledging Power Relations

The lesson study elective course was conceived as a means of developing student teachers' mathematics knowledge for teaching by offering them the opportunity to teach mathematics and to learn from the process of jointly preparing for and reflecting on children's responses during mathematics lessons. As such, it constituted a community of practice among the many others to which the students and I belonged. Wenger (1998) outlines two types of connections which exist between practices: inanimate *boundary objects* (discussed later) and living, vibrant acts of *brokering*. I adopted his metaphor of brokering for my role as course co-ordinator, Knowledgeable Other and researcher of the lesson study community of practice.

> *The job of brokering is complex.* It involves processes of translation, coordination, and alignment between perspectives. It requires enough legitimacy to influence the development of a practice, mobilize attention, and address conflicting interests. It also requires the ability to link practices by facilitating transactions between them, and to cause learning by introducing into a practice elements of another. Towards this end, brokering provides a participative connection—not because reification is not involved, but because what brokers press into service to connect practices is their experience of multimembership and the possibilities for negotiation inherent in participation. (p. 109)

By italicising Wenger's first sentence, I place emphasis on the inherent complexity of the role of convenor of any lesson study group. The complexity of my role was also overtly powerful because these were my students. For ethical reasons, I set about ameliorating this particular power differential by divesting myself of all connection with grading the students. This was possible because I was on study leave for the academic year in which I introduced the lesson study elective course and the ethics of not grading the students related to my role as researcher of the lesson study elective course. The course assignments consisted of a reflective essay summarising learning from the lesson study course and a collaboratively planned mathematics scheme of work for one term for a particular grade level. My two colleague mathematics teacher educators marked the assignments. However, there were still vestiges of power issues for students regarding compliance with requirements for attendance, participation in activities and confidentiality. I feared that an imbalance in power relations between the students and me would be counterproductive to the enterprise and sought to underline from the outset our joint roles as co-researchers (Wagner 1997).

My Role as Broker in the Enterprise

I was careful to articulate my gratitude to the students for participating in the research. Within the constraints of commitments to college and schools' timetables, I assured their freedom to choose which lessons to teach and whom to observe at which site. Freedom to attend and to act was to a certain degree negotiable within the community of practice, but students were aware that the bounds of feasibility were set for all of us by wider and more powerful structures. For example, my authority as course organiser was offset by my dependence on student co-operation with the research. My apparent 'vulnerability' within the System may have acted as an aid to community building. During lesson study cycle two, I was asked to comment on the meaning of the Wagner (ibid) reading which I did by articulating:

Tutor: It says we have to be co-researchers and that's difficult because we have to share power and it's difficult, for you to say…that's important, that's why I'll go with whatever you want with regard to a third and a fifth [class]. OK? (Lesson study cycle 2_Preparation session 1)

This was a poor attempt on my part at initiating a conversation about power relations. That it was not questioned further is perhaps a strong indicator that much of the usual unequal power distribution between students and lecturer obtained in the community of practice, at least at first. On that occasion, I was conscious of my power—by which I mean the measure of my ability to control the environment around me, including the behaviour of other persons—as I commented on reflective journals from the early weeks.

Finola's Research Lesson on Weight

During Finola's research lesson on weight (cycle one), she was disconcerted by a boy's response to her introductory question:

Finola: Your school bag should be…it should be about three and a half kilograms…. That's about one tenth of your body weight. It shouldn't be any heavier than that. Do you think that's…how heavy do you think? …mm…Do you think that means your school bag should be heavy or what weight your schoolbag should be?
Child: It should be only about a quarter of your weight.
Finola: One tenth of your body weight. OK, a quarter of your body weight. OK…but we'll see…because that's what we have to think about today.

This exchange shows how Finola's problem engaged this boy, and had she grasped the Contingency opportunity his question raised she might have questioned his understanding of a quarter of his body weight. Instead, Finola stayed with her prepared lesson script and asked other children: "So how heavy do you think you are and how heavy then is your school bag?" At this stage of the morning, school bags were virtually empty and so when children lifted their bags, some could swing the empty bags on a single finger, and there was a chorus of responses like:

Child: It's only a small bit.
Child: It's light.

Finola appeared unaware of the significance to weight of full and empty school bags and this might be coded either as a Foundation (*awareness of purpose*) or Transformation (*choice of examples*) issue in the lesson, since it had considerable bearing on the weights, which were to be discussed later. In the post-lesson review discussion of both the first child's "quarter of your weight" conjecture and the contingency of the empty—and consequently light—school bags contributed to learning opportunities for the lesson study group who discussed alternative teacher moves.

In her personal journal, Finola had regretted her "inability" and "lack of confidence to deal with the children's observations and misunderstandings." I suggested that the aim of the community of practice was "to build our confidence, so that we can deal with the questions they [children] bring up and its OK not to necessarily know the answer." This confidence-building aspiration was an attempt to empower these prospective teachers, in the hope that they might exercise agency. In retrospect, this was evidence of "deficit-model thinking" on my part (Gutiérrez and Rogoff 2003, p. 19) where I saw myself as bolstering the confidence of the group. Such 'bolstering' was unnecessary. Even at that first reflection meeting, following research lessons in cycle one, there was evidence of a growing self-confidence in relation to mathematical matters, in the discussions about lessons taught.

Ethna's Research Lesson on Fractions

Nóirín's journal entry following the review session on Ethna's lesson (cycle two) is illustrative. This lesson was based on a "fractions with cookies" lesson (Burns 1987).

> Another point we discussed with regards to Ethna's lesson was whether or not to show the class "half a cookie" and a "quarter of a cookie" before they began figuring out the questions for themselves. I think it would be better if, for this lesson, Ethna didn't do this beforehand. We thought that it might take away from them trying to figure it out for themselves, almost giving them the answers before they had begun and leading them in a particular direction. It also might have been easier to assess what they really understood with regard to fractions e.g. what "equal" actually means, the concept of quarters and halves and so on if she had just let them do the activity their own way.

In that instance Ethna had modelled how to divide the cookies and talked the children through the process before she let them do it. The pros and cons of this teaching strategy were discussed during the post-lesson review session. On reflecting on Nóirín and her colleagues' comments, Ethna (Journal 6) defended her decision:

> I felt that because of the nature of the lesson I needed to spend a lot of time on the introduction of the lesson explaining what the children had to do…I did pose a problem for the children to think about and discuss as a whole class and this helped them both think about equally sharing cookies and also gave them an idea of what they had to do for the other activities…The children were aware of the purpose of the exercise and I felt that it was important that I used the right terminology and language for the children that they could relate to and to help them understand the concepts, i.e., share equally, how much does each person get? Does everyone get the same amount?

Participation in lesson study seemed to offer participants a measured chorography of lesson preparation, enactment of teaching and reflection processes where students moved repeatedly between collaborative and individual learning over the duration of the course. I began to realise that the ongoing activity of the community itself was becoming the source of power and the seat of learning.

Nexus of Power Distribution

In broad terms, there were five aspects to my role as broker and varying degrees of power associated with each of them. In an effort to make these more visible, I examine the junctures at which they appeared to interact with student teachers' roles, by first listing the different roles I was aware of assuming throughout the elective course. I was college lecturer/researcher/data manager. I was also lesson study co-ordinator/audio-visual technician and manager. I was Knowledgeable Other. I sought to influence students' attitudes to mathematics and mathematics teaching by offering them opportunities to air and discuss our stories about learning mathematics. This aspect of my role focused on 'affect' in relation to learning mathematics, and was welcomed by the group as an opportunity to strengthen the *identity of participation* of the members (Wenger 1998, p. 215). At least one journal entry specifically mentioned the group ambience as favourable:

> I enjoyed this session. The atmosphere within the group is good and I believe that we will work well together and our understanding of maths will be enhanced. It was good to know that others in the group have a 'fear' of maths and are somewhat uncomfortable teaching it. (Journal 1_Bríd)

In contrast to this 'fear' of mathematics, I had a role within the group as the sole arbiter of 'good' mathematics teaching. I also had a role as the source of challenging mathematics tasks for the 'doing mathematics' aspect of the lesson study course. Contributing to this matrix of roles, the prospective student participants each got an opportunity to assume the role of 'class teacher' on one occasion, each reported on the two lessons they observed and all of them wrote reflective journals on a regular basis. I look first for evidence of power differentials at the interfaces where these various roles obviously interacted.

I am confident that my dependence on student participation for data generation, together with their embracing of the co-researcher role led to interplay of powerful roles and an enhanced mathematics teacher identity for all participants. In relation to executing the technological aspects of the project, the power differential soon swung in the students' favour because of their proficiency with audiovisual technology, as opposed to mine. The remaining roles I assumed appear more problematic and it is here that I perceive the first power imbalance that could contribute to negative experiences for prospective teacher participants.

The act of journaling on the student teacher participants' part was intended to be helpful to them by fostering reflection on their own learning in the lesson study process. It also gave me a window into their emerging identities as teachers of mathematics and while I gave feedback on journals when asked for it, I did not share my field notes with the community. Presumably, it did not occur to anyone to ask for them nor did it occur to me to offer. In retrospect, the "unequal project ownership" identified by Soliman (2001) permeated this aspect of the community enterprise, since I had suggested journaling as an addendum to the lesson study process. The act of journaling complemented the therapeutic aspect of the community of practice and was conceived by me as being beneficial to students. Hannula et al. (2007) theorise the ethics of a therapeutic approach to changing prospective teachers as a duality of agency; *empowerment of students* acknowledges the student teachers' agentic power and *occasioning change* recognises the agency of intervention on the part of the researcher (ibid, p. 156). There is evidence in the data of shifts in personal power associated with both the acts of journaling and participation in bibliotherapy, which attest to a growth in agency among the prospective teachers.

The research lesson is pivotal in any cycle of lesson study and it was in the context of the later research lessons that most evidence of power distribution can be garnered. (See Table 2 for descriptive synopses in cycles two and three). The same two schools—St Peter's serving a middle-class community and St Paul's serving a community of low socio-economic status—were used on each occasion, but these were all 'dive-in' lessons occurring at irregular intervals during the school year, where the student teachers merely had 'guest' status and knew neither the teachers nor the pupils. The act of teaching a lesson is premised on being able to enact 'good' mathematics teaching, as are the acts of observing and reporting on it, so it is especially in these enactments that power was shared and mathematics teacher identities were crafted. However, there was a great deal of power accruing to my role as 'arbiter of good mathematics teaching', and tensions surfaced around it for me, as I questioned my own entitlement to such value judgements. There was a noticeable shift in acceptance of responsibility for considered decisions in planning and follow-through in observing children's responses to the mathematics presented.

Lesson Study Cycles Two and Three

Reviewing the lessons taught in lesson study cycle two was a different experience from that in the first cycle. Both Bríd and Ethna were 'old-timers'—they had participated in tier two and had seen video recordings of their own teaching before—and except for a few whispered asides and some embarrassed laughter they set a tone of serious concentration on following the course of each lesson with a reflective gaze. The DVD was stopped only four times during a first viewing of Bríd's lesson. First, to discuss a Connection opportunity to establish equivalence between fractions; second, to discuss a side issue where Treasa had asked a group of children if one would get a bigger or smaller slice when a pizza was divided among ten people instead of eight. Bríd had planned to ask this but 'forgot'. The third time was an interesting Contingency incident where in dividing pizzas between ten people, a child had tried halves, quarters and then tenths. In order to divide it into tenths, the child realised that the pizza had to be 'reconstituted' first in the absence of a common denominator between fourths and tenths. Treasa requested the fourth stop, where Bríd was asking for explanations of equivalence between six eighths and three quarters. "Could you have drawn it there? You introduced something totally new." The "something totally new" was a Contingent moment where a boy explained the equivalence between 6/8 and 3/4 in terms of dividing six by two to get three and eight by two to get four. The group discussion of Bríd's lesson concluded on an affirmative note, with plans for improvement of future teaching.

Bríd: I was very hesitant about doing the lesson I suppose at the start.
Treasa: And you're more confident now.
Bríd: Because... I don't have the confidence.
Ethna: You were kinda thinking [the mathematics] through yourself.
Bríd: Yeah, I was actually learning from them.
Ethna: And the next time, then the next time you'll probably feel more confident.
Bríd: Next time I'd try to change my presentation, my board work. (Lesson study cycle 2_Review session 1)

In contrast with Bríd, Ethna chose a younger class, which she perceived would be less challenging mathematically. Burns (1987, p. 40) describes the mathematical potential in the summarizing of the lesson, and Ethna's careful preparation of hands-on activity manipulating the fraction pieces on worksheets ensured that she too had food for mathematical discussion with the pupils. Ethna's lesson on fractions was also reviewed at this session, and the issues raised by the student teachers were largely to do with teaching for equity in St Paul's (Boaler 2008). When confronted with digital records of teaching, it appears that the student teachers were drawn into reliving the event and were less analytical in terms of the KQ than they were when reflecting personally at a remove from the lesson. There are resonances here with Schön's (1983) "reflection-in-action" and "reflection-on-action". While video-stimulated recall was also an opportunity for reflection-on-action, the immediacy of the medium, viewed in the presence of community members, was a powerful learning opportunity for reflection-in-action—less structured than the KQ framework.

Two further incidents shed light on this otherwise subterranean vein of power—the right to be an arbiter of 'good' mathematics teaching. I was very aware of the power differential when I was 'pushing/pressing' Róisín as to how she might have used the Contingency opportunities that arose in her lesson on division in cycle three (see Table 2). The extent to which I felt free to do so and the robust manner in which Róisín discussed the pros and cons of reinforcing children's recall and retrieval of number facts in the context of her lesson on division are evidence of "a dispersed and shifting nexus of social power" (Haywood and Mac an Ghaill 1997, p. 268) and a welcome outcome of the lesson study process. There was evidence, too, of a shift or a distribution of power in relation to my role as 'arbiter of good mathematics teaching' in the group's discussion of Nóirín's greater focus on having children articulate their thinking than on enhancing their ability to solve fraction problems. Since the goal of the practice was that student teachers would all become mathematics-teaching practitioners and by implication 'arbiters of good mathematics teaching' then that interface between my role and student participants' roles was extremely important for the practice.

In cycle three there is evidence that a growing sensitivity to context was emerging among the group. A journal entry from Treasa, the same person who, two lesson study cycles earlier, had evaluated her own lesson in St Peter's in terms of 'performance,' commented on Nóirín's lesson:

> I think the children in St Paul's would have benefited from having the circles drawn on the worksheet. The lesson did give a good idea of where the children's weaknesses lie i.e. in terms of knowing what the fractions are called, how to divide equally, equivalence etc. However, I think it is important to note that when prompted, the children, some of them were able to come up with relevant answers so just going on their worksheets and what they did on the board is not a totally accurate representation of their actual knowledge. It was a question of activating their prior knowledge. I think if Nóirín had probed a bit more when going through the first problem she may have found that they could figure out how to do it. (Journal 7_Treasa)

Enactment of teaching problem-based mathematics lessons is advocated by the Irish primary mathematics curriculum (Government of Ireland 1999a) and was normalised by the lesson study community of practice. Engagement in the teaching and observing of these lessons were both powerful experiences for the students.

Despite the growing sophistication evident in the prospective teacher participants' choosing of mathematical tasks for lessons to be taught, the work of providing mathematical tasks for the community to pursue always fell to me. Convinced of the importance of 'doing' mathematics as an essential part of preparing to teach mathematics I chose the mathematics to make a point, that it is easier to develop early fraction concepts if you work in a realistic context. We began with a linear context for division of a whole number by a fraction adapted from the Tsubota (2006) site (see Fig. 1 below).

The division algorithm yielded an answer of seven and a half. By reasoning on a linear diagram, seven pieces of ribbon could be cut away leaving a third of a metre unused. The real task for the students was to rationalise why there appeared to be two different answers to the problem depending on whether they used subtraction or division to solve it and we discussed the meaning of division. Then, we engaged in paper folding relating to the addition and subtraction of fractions from the *Teacher Guidelines* (Government of Ireland 1999b). My goal here was to emphasise for students the value of asking key questions as prompts to encourage student thinking. Finally, we did a 'fun' fraction activity, a game intended to reinforce understanding of fraction concepts (Burns 1996, p. 125). I was more conscious in this cycle of lesson study of my engagement with different contexts: that of learning mathematics,

Here is a 5-metre ribbon. If we cut it into 2/3 m strips, how many strips can we get? How long (m) is the remainder?

Fig. 1 Measurement Division with remainder

that of learning about teaching mathematics and that of learning about teaching student teachers of mathematics. Communication was free and easy and essentially shared by the community during the learning mathematics session, and participants ventilated their relationships with fractions and experiences of learning decontextualised mathematics generally.

Conclusion: Participant Agency and Communal Spaces

I have relied on Wenger's concepts *participation* and *reification* to describe how learning to teach mathematics might have occurred in the lesson study community of practice. I am conscious of marked changes in how these student teachers approached planning for, and teaching of, mathematics. I saw evidence in the student teachers of a growing awareness of the depth and connectedness of mathematical ideas. I traced the development of a much more focused 'eye' on how children build mathematical thinking and witnessed them expend considerable energy in designing opportunities for children to do so. The participants in this study all grew in self-confidence, a self-confidence that recognises personal agency and thrives on communal support. These students and I came to view mathematical knowledge for teaching primary mathematics in new ways and, in consequence, think differently about how mathematical knowledge can be developed or stifled by classroom experiences. Engagement in the community, which practised lesson study, to learn about mathematics teaching engendered enhanced and powerful mathematics teacher identities in all the participants.

References

Boaler, J. (2008). Promoting 'relational equity' and high mathematics achievement through an innovative mixed-ability approach. *British Educational Research Journal, 34*(2), 167–194.

Brown, T, & McNamara, O. (2005). *New teacher identity and regulative government: The discursive formation of primary mathematics teacher education*. London: Springer.

Burns, M. (1987). Fractions with "Cookies". In *A collection of math lessons: from grades 3 through 6*. Sousalito: Math Solutions.

Burns, M. (1996). *50 problem-solving lessons: Grades 1–6*. Sousalito: Math Solutions.

Corcoran, D. (2008). *Developing mathematical knowledge for teaching: A three-tiered study of Irish pre-service primary teachers*. Unpublished doctoral dissertation, University of Cambridge.

Fernandez, C., Cannon, J., & Chokshi, S. (2003). A US-Japan lesson study collaborative reveals critical lenses for examining practice. *Teaching and Teacher Education, 19*, 171–185.

Fernández, M. (2005). Exploring "lesson study" in teacher preparation. In H. L. Chick & J. L. Vincent (Eds.), *Proceedings of the 29th Conference of the International Group for the Psychology of Mathematics Education* (Vol. 2, 305–312). Melbourne: PME.

Government of Ireland. (1999a). *Primary school mathematics curriculum*. Dublin: Stationery Office.

Government of Ireland. (1999b). *Mathematics Teacher Guidelines*. Dublin: Stationery Office.

Gutiérrez, K., & Rogoff, B. (2003). Cultural ways of learning: Individual traits or repertoires of practice. *Educational Researcher, 32*, 19–25.

Hannula, M., Liljedahl, P., Kassila, R., & Rösken, B. (2007). Researching relief of mathematics anxiety mong pre-service elementary school teachers. In J. Woo, H. Lew, K. Park, & D. Seo (Eds.), *Proceedings of the 31st Conference of the International Group for the Psychology of Mathematics Education* (Vol. 1, 153–157). Seoul: PME.

Haywood, C., & Mac an Ghaill, M. (1997). Materialism and deconstructivism: Education and the epistemology of identity. *Cambridge Journal of Education, 27*(2), 261–272.

Lave, J., & Wenger, E. (1991). *Situated learning: Legitimate peripheral participation.* Cambridge: Cambridge University Press.

Lesh, R., & Clarke, D. (2000). Formulating operational definitions of desired outcomes of instruction in mathematics and science education. In A. E. Kelly & R. Lesh (Eds.), *Handbook of research design in mathematics and science education* (pp. 113–149). New Jersey: Lawrence Erlbaum Associates.

Lewis, C., & Tsuchida, I. (1998). A lesson is like a swiftly flowing river: Research lessons and the improvement of Japanese education. *American Educator, 22*(4), 14–17 and 50–52.

Lewis, C., Perry, R., & Murata, A. (2006). How should research contribute to instructional improvement? The case of lesson study. *Educational Researcher, 35*(3), 3–14.

Mills College Lesson Study Group (2000). *Can you lift 100 kg?* [DVD]

National Numeracy Strategy (NNS). (2006). Unit plans year four, measures, available at: http://www.standards.dfes.gov.uk/primary/teachingresources/mathematics/nns_unit_plans/year4/Y$T"Unit4measures/nns_unitplan050703y4t2unit4.pdf

Rowland, T., Huckstep, P., & Thwaites, A. (2005). Elementary teachers' mathematics subject knowledge: The knowledge quartet and the case of Naomi. *Journal of Mathematics Teacher Education, 8*(3), 255–281.

Schön, D. (1983). *The reflective practitioner.* New York: Basic Books.

Soliman, I. (2001). Collaboration and the negotiation of power. *Asia-Pacific Journal of Teacher Education, 29*(3), 219–234.

Sugawara, Y. (2008). Improving mathematics lesson study: why Japanese teachers use dart games for the introduction of area learning. Paper presented at ICME 11. http://www.michinoku.ne.jp/~sugayuki/home1.html=

Tsubota, K. (2006). Excellent Japanese teachers and their works (2): Kozo Tsubota and his work. http://www.michinoku.ne.jp/~sugayuki/tsubota.htm

Wagner, J. (1997). The unavoidable intervention of educational research: A framework for reconsidering research-practitioner co-operation. *Educational Researcher, 26*(7), 13–22.

Wenger, E. (1998). *Communities of practice.* Cambridge: Cambridge University Press.

Preparing for Lesson Study: Tools for Success

Mary Pat Sjostrom and Melfried Olson

One effect of the No Child Left Behind (NCLB) legislation in the United States is that when students do not make progress according to the goals defined by state standards, the school is rated as "needs improvement" and eventually faces a process known as "restructuring." When confronted with declining test scores in mathematics, one elementary school in Hawai'i proactively sought the help of university professors through a Mathematics and Science Partnership grant (MSP). Funded for the period 2005–2008, the grant involved all of the teachers, administrators, and instructional support staff at the elementary school working with three university professors. The goal was to raise student achievement levels by (a) strengthening teachers' mathematical content knowledge, (b) providing teachers with opportunities to reflect on and refine instructional practices, and (c) building a professional community in which teachers collaborate to effect change.

As measured by student achievement data, the partnership was highly successful; student scores on the state assessment improved during the project. Equally important was the growth of all stakeholders (teachers, administrators, and university professors). The external evaluator for the MSP noted that "while there was some fluctuation and attrition throughout the 3 years in roles at the school level, all partners made adjustments and accommodated unanticipated changes" (Uehara 2008, p. 9), as they learned to work together and modify the professional development plan to better serve the needs of the teachers and students. Although Lesson Study was not part of the original plan, it became the focus of year three, leading teachers to comment "on the responsiveness of the higher education partners in meeting their classroom needs" (Uehara 2008, p. 9). The purpose of this chapter is to examine the way in which two years of professional development paved the way for successful implementation of Lesson Study and helped to overcome some of the difficulties commonly encountered when Lesson Study is implemented.

M. P. Sjostrom (✉)
Chaminade University of Honolulu, Honolulu, USA
e-mail: msjostro@chaminade.edu

L. C. Hart et al. (eds.), *Lesson Study Research and Practice in Mathematics Education*,
DOI 10.1007/978-90-481-9941-9_21,

Setting the Stage

The school serves a neighborhood in which the median income is about 25% below the state median. Approximately two-thirds of the students receive free or reduced lunches and approximately 25% of the students are identified as limited English proficiency. Led by an administrator who encouraged and empowered teachers to seek solutions to the problems the school faced, in 2004 the teachers adopted a reform mathematics curriculum. The publisher provided consultants who conducted a series of short-term professional development activities designed to aid in the implementation of this particular curriculum. When identified as a school that *needs improvement* in mathematics and the possibility of restructuring loomed, the school sought an approach that would have long-lasting results. In collaboration with one of the authors, the school sought and received a Mathematics and Science Partnership grant that was funded for three years beginning in August 2005. Before applying for the grant, every teacher in the school agreed to participate. Three professors of mathematics and mathematics education from local universities helped plan and deliver the professional development.

Professional Development in the First Two Years

Seminars in Mathematics Content and Problem Solving

During the first two years, seminars were held at the school every two weeks for approximately two hours at the end of the school day. All teachers and administrators attended these professional development sessions, which focused on reading and writing mathematics. Topics were selected from the content knowledge expected of elementary students as given in the state standards, for example properties of whole number operations. Collaborative problem solving was used to create a professional learning community in which teachers learned to listen to one another, to focus on the process of doing mathematics, and to trust one another. During the second year, teachers also brought student work to the collaborative seminars. They shared and discussed classroom assessment results, learning to analyze student thinking more deeply. Work in the seminars helped the teachers focus on conceptual understanding rather than just procedural knowledge.

Thinker–Doer Activity

A weeklong summer institute was held in June 2006 for an initial cadre of teachers; this workshop was repeated in June 2007 for the remaining teachers. In the summer institute, teachers were introduced to *Thinker–Doer* (Hart et al. 2004), a problem-solving activity in which one participant (Doer) attempts to solve an unfamiliar problem while the other (Thinker) acts as a facilitator. The Doer is asked to think aloud, while the Thinker is instructed to ask questions that encourage the thinking process, without giving away the solution or even affirming the Doer's thinking. The role of Doer gave the teachers in the institute practice in problem solving and communicating mathematically; teachers showed increasing confidence as they repeated the activity during the week. When they took the role of Thinker, teachers learned to listen carefully and to ask questions that promoted flexible thinking and metacognition. While Thinkers reported that it was difficult not to "help" the Doer by giving hints or telling the answer, they grew more comfortable with the role with practice, and learned to listen carefully to understand the problem solver's thinking.

The Thinker–Doer activity related well to the Reflective Teaching Model (RTM) (Hart et al. 2004), which was introduced during the summer institutes and implemented in years two and three.

Reflective Teaching Model (RTM)

Teacher reflection is an integral component in changing teachers' beliefs and practices. Cooney and Shealy (1997) stress the role of reflection in enabling teachers to question traditional methods of teaching and to examine the classroom from different perspectives. In order to effect change in the classroom, a professional development model must include specific methods to support teachers as they plan and teach lessons that engage students in constructing mathematical knowledge. The RTM (Hart et al. 2004) has been shown to support teacher change with positive and lasting effects. The model is grounded in the theories of constructivism and metacognition, and is based on a belief in shared authority; that is, teachers and students are both seen as learners and as teachers. In the RTM teachers experience learning mathematics in a constructivist learning environment. In addition to reflecting on their learning experiences, teachers learn to employ metacognitive strategies as they monitor and regulate their teaching during a lesson, and reflect on the lesson after it is taught (Hart et al. 1992).

In the RTM a pair of teachers work together, with one (Teacher) planning a lesson and the other (Facilitator) facilitating the planning process. As in Thinker–Doer, the Facilitator's role is not to evaluate or assist in the planning, but rather to ask questions to encourage the development of metacognition, with the goal of supporting high quality planning and implementation of the lesson. As the Teacher plans, the Facilitator asks questions such as these:

1. How do you usually teach this lesson? (thinking about prior experiences)
2. What do you expect students to already know about this concept? (thinking about student prior experiences)
3. What should students know by the end of this lesson? (setting explicit goals)
4. What misconceptions might students have? (preparing for possible interventions)
5. How do you expect students to respond? (anticipating student responses)

As the Teacher teaches the lesson, the Facilitator observes (videotaping if live observation is not possible), not for the purpose of evaluating but simply to take descriptive notes. A debriefing session follows, in which the Facilitator asks questions to encourage the Teacher to reflect on his or her instructional decision-making and the effectiveness of the lesson:

1. How do you think the lesson went? (analyzing and reflecting in general)
2. What worked? What did not work? (analyzing and reflecting more specifically)
3. What would you do differently if you could teach the lesson again? (revising)

During the debriefing the focus is on increasing the Teacher's awareness of instructional decisions, the impact of the decisions on students, and how they assisted or hindered the achievement of instructional goals.

The RTM encourages teachers to view teaching as a problem-solving activity, to think heuristically rather than algorithmically. Rather than scripting a lesson, the Teacher develops a set of strategies to use as the teaching solution for this lesson. As in problem solving, the solution process may be modified as the lesson progresses. The power of this process is illuminated by the reflection provided by one teacher:

> I have never taught this lesson before so I was planning to follow the plan in the [teacher's guide]... During the sessions I began to rethink my plans. Briefly talked to [professors]... with the Teacher's Lesson Guide and new insights from [professors], I decided to completely rewrite my lesson plan. At first, I panicked and felt overwhelmed and confused.
>
> I went back to my objective... and eliminated parts of the lesson that did not work toward that objective. In doing this, it also eliminated some problems I anticipated.
>
> I thought about learning modalities and decided to add kinesthetic, visual and computational activities.
>
> I thought more deeply about management of materials, types of materials to be used, classroom arrangement, timing, instructional time vs. student work time.
>
> One thought kept leading to more thoughts about this lesson...(personal communication, March 2008).

Year Three: Lesson Study

Some improvement in student achievement, as measured by standardized testing, was observed during the first two years of the grant. Teachers were particularly enthusiastic about the RTM, but felt that the content-knowledge seminars planned by the professors did not address their needs in the same way. Entering the third and final year of the project, the teachers, empowered by the RTM, Thinker–Doer, and other components of previous years, took more responsibility for determining the direction of the professional development work. Lesson Study was suggested by a team member, and through the shared authority established by the teachers through participating in the RTM, the idea was soon accepted by the leadership team and the teachers.

Several principles of Lesson Study were used as a basis for the professional development in year three: (a) expect improvement to be continual, gradual, and incremental; (b) maintain a constant focus on student learning goals; (c) focus on teaching, not teachers; (d) make improvements on teaching in context; (e) provide opportunities for teachers to collaborate as they plan instruction and analyze teaching; and (f) build a system in which teachers can learn from their experiences (Stigler and Hiebert 1999).

To prepare for the implementation of Lesson Study, ideas from Watanabe (2002) and Stigler and Hiebert (1999) were compared and contrasted with prior work using the RTM. While the RTM in many ways prepared teachers for Lesson Study, in other ways it added to the confusion. The RTM focuses on the reflective practice of the teacher, whereas in Lesson Study the focus is on student *learning* rather than the teacher who implements the lesson. The RTM cycle of plan-teach-debrief takes one or two days, and may be repeated in any time frame. The cycle used in Lesson Study to develop and refine a research lesson takes several months. However, in both practices the goal is to improve teaching and thus improve learning, and teachers take primary responsibility through a cycle of planning, teaching, observing, and debriefing. Table 1 provides a comparison of the RTM and Lesson Study.

Table 1 Comparison of Reflective Teaching Model and Lesson Study

	Reflective Teaching Model	Lesson Study
Planning the lesson	• Planning is the responsibility of the teacher; partner facilitates reflection • Lesson is usually designed for a day in the near future • Lesson may be based on a textbook lesson that needs some modification or enhancement • Planning takes about an hour	• Collaborative—lesson is "owned" by the group • Problem may be general (e.g., interest students in math) or specific (e.g., improve student understanding of adding fractions with unlike denominators) • Planning may take several months
Teaching	• The teacher teaches the lesson he/she planned • Plans may be modified during lesson in response to student responses • Facilitator observes lesson and takes note of critical aspects to be reviewed during debriefing	• One teacher in the group teaches the lesson, but all participate in preparation • Group observes lesson • Observers walk around during lesson, observing and taking notes on what students are doing
Debriefing	• The teacher and facilitator debrief soon after the lesson is taught • Tone is non-evaluative. Purpose is to focus the teacher's thinking on significant aspects of the lesson, particularly instructional decisions and student responses • Focus is on increasing teacher's awareness of instructional decisions and their impact on students	• Group debriefs soon after the lesson (often that afternoon) • All members speak about how the lesson worked, parts that were problematic • Focus is on lesson, not the teacher • Revision by group; may decide to change materials, activities, problems posed, etc. • Revised lesson is retaught
Product	• Repeated involvement in the process develops metacognition • Teacher becomes aware of instructional choices and decisions • The result is a more flexible and open learning environment in that teacher's classroom	• Product is a "research lesson" that can be shared with others • Teachers learn to think deeply about the way the experiences they structure in their classrooms will facilitate students' understanding of mathematics

There were not enough teachers to make a Lesson Study team for each grade level and there was a desire that every teacher be involved in a team. Three teams were organized, grades preK-1, 2–3, and 4–5. Each team included administrative and support staff, as well as a professor to facilitate the process. Each team set its own agenda, selecting and researching the topic, planning and teaching a lesson, and analyzing the results. During the time period from September to April, two of the teams were able to complete an entire cycle; one of the teams revised and retaught the research lesson.

A description of the effort of the grade 4–5 team is illustrative of the process employed. The topic of equivalent fractions was chosen because it was of interest to teachers at both grade levels and they found the concept hard to teach and hard for students to learn. The team examined prior student data and was surprised to find that while 4th grade students did not do well with equivalent fractions, 5th grade students did. Work samples collected from the 5th grade students confirmed their understanding. Based on this data, the team decided to make grade 4 the focus of the Lesson Study. At the same time, they gathered student data; team members examined research on rational numbers and proportional thinking (Lamon 2007). They were guided by a statement from Van de Walle (2007): "All students should eventually be able to write an equivalent fraction for a given fraction. At the same time, the rules should never be taught or used until the students understand what the result means." (p. 308). After examining the context of equivalent fractions in the curriculum, a decision was made to focus on the lesson *Many Names for Fractions*. The lesson was developed between August 15 and December 8, planned in six organized sessions of 135 minutes each, plus time spent by individual teachers between sessions. The lesson was taught on December 12, 2007.

The lesson developed by the team focused on having children learn about equivalent fractions through the use of cards that had regions shaded green or white. All cards were the same-sized rectangle with partitions that could represent a fraction. For example, two-thirds was represented as two regions shaded green and one shaded white. Other fractions equivalent to two-thirds had four regions shaded green and two white, six shaded green and three white, or eight shaded green and four white. The lesson outline included a plan for the teacher's words and actions, specific activities in which the students would be engaged, and expected student responses. Prior to the lesson, the team decided on a structure that would allow them to document the actual words and actions of the teacher and responses of the students. The lesson, taught by one of the 4th grade teachers, was videotaped with three other team members in attendance. All team members participated in debriefing the lesson, viewing the video with printed copies of the lesson in hand to compare the implemented lesson to the intended lesson. The debriefing compared anticipated student responses to actual student responses.

Several times during the debriefing, the students' work and responses to the teacher's questions provided the impetus for a good discussion. In one case, children were asked to find cards that represented one-half. One student presented six-twelfths and the teacher asked, "How could you prove that?" It was expected that the student would show that the same amount of green was shaded in one-half and six-twelfths. Instead, the child replied "6 is half of 12." At this point, the observers had a lengthy discussion of "What makes something one-half?" Clearly the team expected a visual response because they had chosen visual materials with which to work, but the child used number relationships without verbal reference to any visual representations. In another instance, a child was given a card with three regions shaded green and nine shaded white. The team had to listen carefully to understand his reasoning as he explained to another student why this card represented the same as one-fourth. He did so without any direct comparison to the card that is usually used to represent one-fourth. He counted "1-2-3, pause; 1-2-3, pause; 1-2-3, pause; 1-2-3, pause" and argued the region usually used to represent three-twelfths was equivalent to one-fourth. The student clustered the three green regions into one larger region and replicated the new region three more times to recognize the card could be seen to represent one-fourth. While this explanation did not convince the other student, the observers reflected on why this was an accurate interpretation, but was clearly not a response anticipated in the planning of the lesson. Based on the comparison of predicted and observed student responses and the discussions of the students' mathematical thinking (as inferred by the teachers), the team made notes for revisions of the lesson.

Summary

Several potential obstacles to Lesson Study have been discussed in the literature and this project encountered many of them. However, the work of the first two years of the MSP effectively minimized the obstacles. Following is a brief description of several obstacles and how the project was able to satisfactorily address them:

1. Teachers in the United States want quick results. This cultural trait is heightened by the pressure for accountability stimulated by NCLB (Chokshi and Fernandez 2004). By the time Lesson Study began in the third year of this project, the teachers were aware of the need for a long-term look at professional development. Teachers recognized that what they do in classrooms has an impact on student learning. Gains in student achievement during the first two years also kept teachers focused on improvements yet to be achieved.
2. A district-level emphasis on fidelity to the adopted curriculum often discourages teachers from straying from the textbook. This attitude is illustrated by one teacher's response to a Lesson Study question. When asked, "What have students had problems with in this curriculum?" the teacher responded "9.3" (referring to a lesson in the textbook). Teachers came to recognize that for two years prior to this project, they had maintained fidelity to the curriculum but student achievement did not improve significantly. The examination of student work across grade levels on similar topics helped focus attention on the impact of teachers' instructional decisions.

3. In the United States, classroom observations "have traditionally been conducted in the context of performance evaluation." (Chokshi and Fernandez 2004) As a result, many teachers feel anxious even when observed by their peers. During year three, this was not an issue for the MSP participants. As a result of the RTM experience during the first two years, teachers were comfortable being observed and did not view the observation as an evaluation.
4. Elementary teachers have little time within the school day to devote to the group planning and observation that Lesson Study involves. Funding built into the grant allowed for time to meet and discuss ideas, visit classrooms, and cooperatively plan.
5. Teachers in the United States, especially elementary teachers, often lack content knowledge in mathematics (Ma 1999). Teachers are likely to focus on procedural knowledge rather than conceptual understanding. Seminars during the first year of the MSP focused on mathematical content. The RTM, Thinker–Doer, and analyses of student work generated opportunities for teachers to move from a focus on procedural knowledge toward conceptual understanding of elementary mathematics.
6. Teachers in the United States are used to working independently and often have difficulty collaborating (Fernandez and Chokshi 2002). Collaborative problem solving and the RTM during the first two years provided a foundation for the professional learning community necessary for work on Lesson Study.

In conclusion, two years of work prior to the initiation of Lesson Study provided a strong foundation upon which to build. The implementation of Lesson Study in year three integrated the three components of the MSP: it strengthened teachers' mathematical content knowledge, provided them with opportunities to reflect on and refine instructional practices, and developed a collaborative community of teachers. As a result, as shown in Fig. 1, student achievement in mathematics increased. In

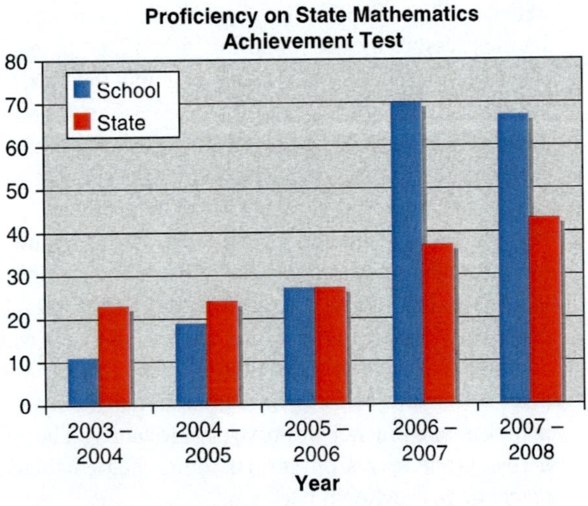

Fig. 1 Percent of students who achieved proficiency on the state mathematics test. (Accountability Resource Center Hawai'i 2008)

fact, in 2008 the school was ranked among the top 10 schools in the state for best average increase in mathematics proficiency scores.

References

Accountability Resource Center Hawai'i. (2008). School accountability: No child left behind. http://arch.k12.hi.us/school/nclb/nclb.html. Accessed 10 April 2009.

Chokshi, S., & Fernandez, C. (2004). Challenges to importing Japanese lesson study: Concerns, misconceptions, and nuances. *Phi Delta Kappan, 85*, 520–525.

Cooney, T. J., & Shealy, B. E. (1997). On understanding the structure of teachers' beliefs and their relationship to change. In E. Fennema & B. S. Nelson (Eds.), *Mathematics teachers in transition* (pp. 87–109). Mahwah: Lawrence Erlbaum Associates.

Fernandez, C., & Chokshi, S. (2002). A practical guide to translating lesson study for a U.S. setting. *Phi Delta Kappan, 84*(2), 128–134.

Hart, L. C., Schultz, K., Najee-ullah, D. H., & Nash, L. (1992). Implementing the professional standards for teaching mathematics: The role of reflection in teaching. *Arithmetic Teacher, 40*(1), 40–42.

Hart, L. C., Schultz, K., & Najee-ullah, D. H. (2004). Thinker/Doer paired problem-solving: A long-term teacher development activity (workshop facilitator's notes). In G. Bright (Ed.), *Professional development resources* (pp. 31–44). Reston: National Council of Teachers of Mathematics.

Lamon, S. J. (2007). Rational numbers and proportional reasoning. In F. K. Lester (Ed.), *Second handbook of research on mathematics teaching and learning* (pp. 629–667). Charlotte: Information Age Publishing.

Ma, L. (1999). *Knowing and teaching elementary mathematics: Teachers' understanding of fundamental mathematics in China and the United States.* Mahwah: Lawrence Erlbaum Associates.

Stigler, J. W., & Hiebert, J. (1999). *The teaching gap.* New York: The Free Press.

Uehara, D. (2008). *Excellence in mathematics summative evaluation report.* Unpublished report.

Van de Walle, J. (2007). *Elementary and middle school mathematics: Teaching developmentally* (6th ed.). Boston: Pearson Education, Inc.

Watanabe, T. (2002). Learning from Japanese lesson study. *Educational Leadership, 59*(6), 36–39.

Response to Part V: Ideas for Developing Mathematical Pedagogical Content Knowledge Through Lesson Study

Makoto Yoshida and William C. Jackson

Introduction

Japanese lesson study was introduced to the United States by researchers in the late 1990s as a possible professional development approach to improve the teaching and learning of mathematics in US classrooms (Lewis and Tsuchida 1998; Stigler and Hiebert 1999; Yoshida 1999). Characteristics of lesson study, such as collegial planning, observation and reflection on classroom teaching and learning, and continuous professional development throughout a teacher's career, have been recognized as important ideas (Lewis and Tsuchida 1998; Stigler and Hiebert 1999; Yoshida 1999; Fernandez and Yoshida 2004; Takahashi and Yoshida 2004).

More recently, the National Staff Development Council (NSDC) recognized lesson study as one of the "powerful designs" for developing Professional Learning Communities (Easton 2004). Because of this recognition, and the promotion of lesson study by other national organizations such as the National Council of Teachers of Mathematics (NCTM), more teachers and schools have become interested in lesson study. As more schools and districts consider lesson study in the United States, administrators and teachers often ask questions like, "Does lesson study help teachers to develop the strong content and pedagogical knowledge needed to improve classroom teaching and learning?" and, "Does lesson study help to improve student learning and test scores?"

These questions are not easy to answer because lesson study was introduced in the United States fairly recently and few sites have conducted it continuously, consistently, and systematically (Lewis et al. 2005; Chokshi and Fernandez 2004). In addition, in our opinion, most lesson study groups in the United States and outside of the United States that began doing lesson study within the last decade are still in the learning stage and generally have not yet approached the high level of lesson study often observed in Japan. Because of this reason, rather than rushing to do research on lesson study's impact on student achievement to determine if it

M. Yoshida (✉)
Center for Lesson Study, William Paterson University, Wayne, NJ, USA
e-mail: yoshidam@wpunj.edu

L. C. Hart et al. (eds.), *Lesson Study Research and Practice in Mathematics Education*,
DOI 10.1007/978-90-481-9941-9_22,

is effective for improving test scores, it may be beneficial at the present time to focus on lesson study's impact on teachers, such as how lesson study affects teachers' mathematical content and pedagogical knowledge. Although many people have questions about lesson study's impact on teaching and learning, most agree that what a teacher knows is one of the most important influences on what teachers can do in the classroom to support students' learning.

The previous chapters written by Sjostrom and Olson, Bruce and Ladky, and Corcoran provided research on how lesson study impacted teachers and student teachers, particularly in the area of pedagogical content knowledge. In this chapter we will try to reflect on the current state of lesson study in the United States as it relates to improving teachers' pedagogical content knowledge in mathematics. By identifying the issues surrounding the improvement of teachers' pedagogical content knowledge through lesson study, we will suggest some ideas to improve lesson study.

The Need for Pedagogical Content Knowledge in Mathematics

Many research documents suggest that mathematics teachers in the United States, particularly in elementary and middle schools, do not have sufficient content and pedagogical knowledge in mathematics (Ma 1999). One solution that has been suggested is to require US teachers to take more mathematics courses in preservice programs (PROMISE 2006). However, unless these courses are focused on developing a deeper and well-connected understanding of the mathematics teachers are expected to teach, aspiring teachers may not acquire other important knowledge necessary to improve student learning.

In its landmark book *Adding It Up* (Kilpatrick et al. 2001), the National Research Council (NRC) introduced a vision of mathematical proficiency for students. This document strongly suggests that teachers need to acquire integrated knowledge about the mathematics they are teaching, not just knowledge about mathematics. Teachers must also have knowledge about how students' mathematical understanding is developed and how different instructional methods help to develop different types of mathematical proficiency.

Shulman (1987) outlined the complex set of knowledge that teachers need to teach effectively, including the knowledge of pedagogy, content, and learner cognition.

> The key to distinguishing the knowledge base of teaching lies at the intersection of content and pedagogy, in the capacity of a teacher to *transform* the content knowledge he or she possesses into forms that are pedagogically powerful and yet adaptive to the variations in ability and background presented by the students. (p. 15)

Since teachers combine this complex set of knowledge with their own beliefs in order to facilitate classroom teaching and learning, they must adapt and transform this knowledge in the context of the classroom, the setting in which they work on improving instruction. This kind of practical content knowledge is often referred to as "pedagogical content knowledge" and includes the knowledge of how mathematical topics are related across grade levels, how student understanding is built up, the kinds of representations and manipulatives that support students' learning, students' state of prior learning, common misconceptions, and how various teaching strategies contribute to learning. Although we agree that improving preservice and inservice courses for prospective teachers is important, we believe that they are insufficient to significantly develop the pedagogical content knowledge they need for effective classroom practice.

Japanese mathematics educator Yoshishige Sugiyama (2008) outlined three levels of mathematical instruction and the knowledge teachers need for each one. In the first level, the teacher knows how to perform mathematical procedures and formulas and demonstrate them to the students. This type of teaching is characterized by teacher demonstration followed by student practice. It is not necessary to have pedagogical content knowledge for this kind of teaching, only procedural knowledge. In the second level, the teacher has conceptual understanding of how and why mathematical procedures and formulas work and tries to explain this understanding to students. Teachers at this level may incorporate manipulatives or small group work, but they are still the ones who convey the rules and procedures. A certain degree of pedagogical content knowledge is necessary at this level. In the third level, the teacher not only understands mathematical procedures and formulas and their conceptual underpinnings, but also tries to develop students' autonomy by helping them to discover and create new mathematical ideas. This type of teaching is often characterized by lessons in which students are actively involved in creating their own learning. The teacher's role at this level is to facilitate and nurture students' mathematical thinking and understanding. This type of instruction requires deep pedagogical content knowledge and has been highlighted in research about Japanese mathematics classrooms (Stigler and Hiebert 1999). Furthermore, it has been shown that lesson study plays a significant role in developing this type of teaching in Japan (Fernandez and Yoshida 2004; Takahashi and Yoshida 2004).

Conducting Kyozaikenkyu to Improve Teachers' Knowledge

One of the reasons why lesson study contributes to moving teachers toward the type of student-centered lesson described by Sugiyama is that during lesson study, teachers work together with their colleagues to learn about subject matter, pedagogy, and student thinking and then apply what they learned by developing a lesson plan. Since the plan they develop is implemented in an actual classroom setting, lesson study is a powerful way to enhance teachers' pedagogical content knowledge. In order to make the most of this learning opportunity, however, three essential components are needed: (1) a well-researched lesson plan with a clear rationale; (2) live observation of the lesson with various participants; and (3) a focused post-lesson discussion based on participants' observations. These elements are the heart of lesson study and are essential for teachers to be able to reflect on their own practice and develop lifelong learning skills for improving classroom teaching and learning.

The structure and process of lesson study creates many opportunities for teachers to study and learn with their colleagues. Many researchers point out, however, that although the process of conducting lesson study seems simple on the surface, learning to conduct lesson study effectively is not an easy task (Chokshi and Fernandez 2004; Yoshida 2008). In order for lesson study to be effective, teachers need to develop additional skills such as how to study instructional materials, plan lessons, observe and collect data on student learning, reflect on lessons, and engage in discussion about lessons with colleagues.

Sjostrom and Olson's chapter describes the project's success for improving pedagogical knowledge with a combination of lesson study, content workshops, the Reflective Teaching Model, the Thinker–Doer activity, and analysis of student work. They reported that teachers' knowledge shifted from a procedural based understanding of elementary mathematics to a more conceptual understanding. These subsets of activities that surrounded lesson study enhanced teachers' opportunities to do and solve problems, study instructional materials, discuss and reflect on their thinking, and anticipate student thinking.

Corcoran's chapter described how doing mathematics and solving problems together with group members helped prospective mathematics teachers to be able to plan lessons based on posing realistic problems and focusing on children's responses. She discussed how the prospective teachers' limited understanding of mathematics created a challenge for them to plan student-centered lessons.

Bruce and Ladky shared a similar view by pointing out that the backstage activities between the four simple steps of lesson study—identifying goals, planning a research lesson, implementing and observing, and reflecting and discussing—need to be made more explicit by doing them intentionally. For example, in the lesson planning stage, they point out that "playing" and "figuring out" how to solve problems or use manipulatives were deeply connected to teachers' understanding of mathematical concepts and how those concepts could be represented and investigated by students. Moreover, they suggested that adequate time to develop conceptual understanding was a critical component for the teachers to develop better lessons.

The points made by these researchers suggest that in order for lesson study to be effective and successful, teachers need to have adequate time to study instructional content, materials, and students' anticipated responses during the lesson study process, and that these exercises must be done intentionally and made more explicit.

In our experience, many US teachers often go through the surface features of lesson study but fail to engage in it in a way that sufficiently impacts pedagogical content knowledge because these skills may be lacking or underdeveloped. The problem is exacerbated by the fact that strong pedagogical content knowledge supports good planning, observation, and discussion, and if this is lacking, it is difficult to conduct lesson study in an effective and meaningful way.

We believe that one of the main reasons that many US teachers often fail to conduct lesson study in a way that truly impacts their pedagogical content knowledge is the failure to engage adequately in *kyozaikenkyu* (instructional material investigation). During the *kyozaikenkyu* process, teachers investigate instructional materials such as textbooks, standards, teacher's manuals, manipulatives, research papers, and other resources in order to develop an in-depth understanding of the content they are teaching (Takahashi and Yoshida 2004; Watanabe et al. 2008; Yoshida 2008). In this process, teachers often solve problems and investigate the use of manipulatives in order to anticipate students' thinking, solutions, and strategies, and move the classroom discussion in a way that constructs understanding among students. Lesson study provides an opportunity for teachers to engage in deep *kyozaikenkyu* as they seek to understand mathematics, scope and sequence, and student thinking, as well as explore possible problems, activities, and manipulatives (Watanabe et al. 2008). Since *kyozaikenkyu* is something that teachers must also do for their everyday lessons, lesson study provides opportunities for teachers to conduct *kyozaikenkyu* more effectively. *Kyozaikenkyu* is a key part of Japanese teachers' lesson study practice, but our experience indicates that in the United States this practice is often not done carefully because teachers are not familiar with or accustomed to doing such investigations during the planning process, and insufficient time is available to do so. We believe that improving the *kyozaikenkyu* process is a way to improve teachers' pedagogical content knowledge.

We have observed that when US teachers are given the opportunity to conduct lesson study together with experienced lesson study practitioners, they are able to improve their pedagogical content knowledge. One such group consisted of US and Japanese teachers that collaborated together on a unit on the area of plane figures (Akasaka et al. 2005). During the *kyozaikenkyu* process, they investigated Japanese textbooks and compared them to Singaporean and American textbooks to study how the concept of area is developed. This process allowed them to think about how the knowledge of area is built up and how students' prior knowledge could be tapped to help them find the areas of new, unlearned shapes. It also helped them to think about the kinds of mathematical activities they wanted to provide for their students as well as the merits and value of these activities.

For their research lesson, they chose to develop a lesson on determining the area of a triangle. During their *kyozaikenkyu*, they realized that there are many different kinds of triangles that they could provide students with in order to think about how to find the area but some triangles may be more advantageous than others. They found out, for example, that it is easy to find the area of a right triangle since students have already learned how to find an area of a rectangle, and they could see that the right triangle is a half of the rectangle. Since a right triangle is a very special kind of triangle, however, they realized that it might not be the best choice for generalizing the method to other kinds of triangles since other triangles are not always a half of a rectangle. In addition, finding the area of a more generically shaped triangle might provide a more challenging task for the students and the solution methods they come up with may be more applicable for finding the areas of other shaped triangles. The teachers discussed the advantages and disadvantages of the various triangles and thought about what dimensions would facilitate students' ability to transform the triangle into familiar shapes, such as rectangles or parallelograms. They also cut and transformed the triangles themselves to think about the different methods students may come up with, how to describe the various methods with mathematical expressions, and how to connect the various methods to generalize the formula for the area of a triangle.

The teachers also considered the students' prior learning experiences and their current state of learning to help determine their anticipated reactions and misconceptions when developing the lesson. This enabled them to understand the merits of the mathematical activity they were asking the students to engage in and how the lesson developed other important skills besides learning a formula, such as writing mathematical expressions, engaging in meaningful discussion about mathematics, generating multiple solution methods, and comparing, contrasting, and generalizing the methods. A thorough *kyozaikenkyu* improved the quality of the research process and helped teachers produce a well-thought-out research lesson plan with a clear rationale, clear goals, a clear plan to implement the lesson, clear outcomes, and a clear method for evaluation. Furthermore, by conducting good *kyozaikenkyu* and communicating what the group studied and learned to observers, observation and data collection became more purposeful and the post-lesson discussion more focused and deep.

Engaging in productive *kyozaikenkyu* is not an easy task, but it is necessary to conduct effective lesson study. Although the above example is based on nonformal research, we have observed many other instances where we believe that US teachers have improved their pedagogical content knowledge and moved closer towards Sugiyama's third level of mathematical instruction by engaging in thoughtful and thorough *kyozaikenkyu*. We have also observed instances where *kyozaikenkyu* was either largely ignored or superficial with the result being little if any improvement of pedagogical content knowledge. More formal research needs to be conducted on this topic, but our observations lead us to believe that merely conducting lesson study without serious *kyozaikenkyu* may not improve pedagogical content knowledge.

Using Focused Materials to Improve Kyozaikenkyu

One possible reason why US teachers do not sufficiently engage in the *kyozakenkyu* process may be the unfocused nature of US curricula and textbooks (Schmidt et al. 2002). Documents such as NCTM's *Curriculum Focal Points for Prekindergarten through Grade 8 Mathematics* (National Council of Teachers of Mathematics 2006) suggest that teachers should focus on a small number of significant mathematical topics at each grade level and understand clearly when and how those important concepts are taught and connected in order to develop understanding. These topics were chosen because they help students to carefully build the mathematical knowledge they need to understand more advanced mathematics in the future. Our experience suggests that a possible way to help teachers effectively engage in *kyozaikenkyu* is to provide them with concise curricular materials that are grounded in strong content and pedagogical knowledge during the lesson study process.

One possibility for these materials is textbooks from countries highlighted in the TIMSS, such as Singapore and Japan, which are grounded in strong mathematical content and pedagogical knowledge and have recently become available in the United States. The *kyozaikenkyu* of the group that studied the area of plane figures mentioned previously was facilitated and enhanced by studying Japanese textbooks. Since these textbooks are concise, coherent and focused, it is easier for teachers to see how mathematical ideas are built across grade levels and what important mathematical concepts students should know at each grade level. Unlike typical US textbooks, they do not contain many extras, such as long introductions, colorful pictures, biographical stories, or discussions of nonmathematical topics. They also include clear representations and concise explanations. Furthermore, the Common Core State Standards for Mathematics were released in June 2010 that emphasize a focused and coherent curriculum. In order to understand this new curriculum, we also believe that studying textbooks like those from Japan and Singapore provides in-depth insights about what a focused and coherent curriculum means.

One interesting aspect of Japanese textbooks is that they contain ideas gained from *kyozaienkyu* and lesson study (Takahashi and Yoshida 2004). This is because they are developed by experienced lesson study practitioners (teachers), as well as mathematics educators and mathematicians that are familiar with lesson study. Japanese teachers conduct *kyozaienkyu* during lesson study by studying these textbooks, as well as teacher's manuals, and the Teaching Guide for the Japanese Course of Study, which explains how and why topics are sequenced, important mathematical concepts students should learn at each grade level, and ideas for teaching these concepts. Japanese teacher's manuals provide a clear rationale for the topics and learning sequence, and discuss how students' prior knowledge can be tapped and utilized in each lesson, as well as typical student reactions and misunderstandings (Takahashi et al. 2005; Takahashi and Yoshida 2004). Japanese mathematics textbooks are not only helpful for students to learn mathematics, but also for teachers to learn how to teach mathematics in classrooms, which makes them a useful resource for conducting effective *kyozakenkyu* and lesson study.

We are not suggesting that schools replace their curriculum and textbooks in order to conduct lesson study. After studying focused materials such as those described above, teachers can go back to their own textbooks and teacher's guides and discuss similarities and differences. Since the teachers already see the bigger picture of how mathematical ideas are connected by studying focused and coherent curriculums, they have better and more critical eyes to look at their own textbooks. Needless to say, in order to do *kyozaikenkyu* effectively, teachers need to have adequate time to collaborate and investigate instructional materials.

Accessing Knowledgeable Others to Improve Kyozaikenkyu

Another way to help teachers conduct effective *kyozainenkyu* is by having access to knowledgeable others. Corcoran's chapter discussed the important role of knowledgeable others for effective lesson study. Since novice lesson study practitioners do not have sufficient experience and knowledge to conduct lesson study and *kyozaikenkyu* effectively on their own, it is necessary for someone to facilitate the process by educating and guiding them. Successful and meaningful lesson study occurs when participants feel that that they have not just developed a successful lesson but they have learned pedagogical content knowledge for the topics they studied.

One significant difference between lesson study in Japan and the United States is that Japanese teachers can take advantage of the accumulated experiences and knowledge of Japan's long history of conducting lesson study. Many experienced lesson study practitioners have been nurtured over time, and their significant contribution to the lesson study community improves the quality of lesson study in Japan. In addition, many university professors participate in lesson study as knowledgeable others and work with lesson study practitioners to maintain high quality lesson study. These knowledgeable others are not only knowledgeable about subject matter content or pedagogy, but also experienced observers and discussants of classroom instruction and student learning. Also, some of them have had extended experience as classroom teachers prior to working at universities. Many of them serve as consultants for lesson study groups and work with teachers for extended periods of time. They also know how to support lesson study and collaboratively work with teachers in order to improve student learning and understanding through classroom practice. In this sense, they have more experience working in lesson study communities compared to the possible knowledgeable others we may be able to find in the United States.

The participation of these experienced lesson study practitioners and knowledgeable others is also important to improve the quality of *kyozakikenkyu*. Even if a group of teachers engages in lesson study, if all the teachers have weak knowledge about the content they are studying, the quality of lesson study may remain low. However, if somebody who is knowledgeable about content and pedagogy is involved, the quality of lesson study and *kyozaikenkyu* increases. In English, we say, "Two heads are better than one," and in Japanese, "Three heads together produce *Manjusri*" (Buddhist enlightenment and wisdom).

To improve lesson study in the United States, we believe it is very important to have experienced lesson study practitioners and knowledgeable others to support teachers and ensure the quality of lesson study and *kyozaikenkyu*.

Conclusion

We believe that the issues discussed in this chapter for improving teachers' pedagogical content knowledge through lesson study are important to increase the quality of lesson study and enhance the teaching and learning of mathematics in classrooms, and that further research is warranted. The research of Sjostrom and Olson, Bruce and Ladky, and Cocoran shows that more careful analysis and design of intentional activities in the lesson study process are necessary to have effective and meaningful lesson study that develops strong pedagogical content knowledge in teachers. In addition, we believe these ideas may help educators to engage in a more lively discussion on how to improve lesson study. Obviously, this requires a cultural shift in how teachers, teacher educators, and professional development providers think about professional development, and it will take time to make such a cultural shift. Steady change can happen if schools have a clear, long-term vision of professional growth through lesson study, teachers are willing to engage in productive *kyozai-kenkyu* during the lesson study process, and helpful materials and knowledgeable others are available to support their efforts.

References

Akasaka, J., Ishida, T., Jackson, W., Spain, D., & Toki, R. (2005). *A unit plan on the area of plane figures*. Unpublished document.

Chokshi, S., & Fernandez, C. (2004). Challenges to importing Japanese lesson study: Concerns, misconceptions, and nuances. *Phi Delta Kappan, 85*(7), 520–525.

Common Core State Standards Initiative. (2010). Common core state standards for mathematics. The National Governors Association center for Best Practices and the Council of Chief State School Officers. http://corestandards.org.

Easton, L. B. (Ed.). (2004). *Powerful designs for professional learning*. Oxford: National Staff Development Council.

Fernandez, C., & Yoshida, M. (2004). *Lesson study: A Japanese approach to improving mathematics teaching and learning*. Mahwah: Lawrence Erlbaum Associates.

Kilpatrick, J., Swafford, J., & Findell, B. (Eds.). (2001). *Adding it up: Helping children learn mathematics*. Washington: National Academy Press.

Lewis, C., & Tsuchida, I. (1998). A lesson is like a swiftly flowing river: How research lessons improve Japanese education. *American Educator, 22*(4), 12–17, 50–52.

Lewis, C., Perry, R., & Murata, A. (2005). How should research contribute to instructional improvement? The case of lesson study. *Educational Researcher, 35*(3), 3–14.

Ma, L. (1999). *Knowing and teaching elementary mathematics: Teachers' understanding of fundamental mathematics in China and the United States*. Mahwah: Lawrence Erlbaum Associates.

National Council of Teachers of Mathematics. (2006). *Curriculum focal points for pre-kindergarten through grade 8 mathematics*. Reston: National Council of Teachers of Mathematics.

Promoting Rigorous Outcomes in Mathematics and Science Education (PROMISE). (2006). *Knowing mathematics: What we can learn from teachers* (Research Report Vol. 2). East Lansing: Michigan State University.

Schmidt, W., Houang, R., & Cogan, L. (2002). A coherent curriculum: The case of mathematics. *American Educator, 26*(2), 1–18.

Shulman, L. S. (1987). Knowledge and teaching: Foundations of the new reform. *Harvard Educational Review, 57*(1), 1–22.

Stigler, J., & Hiebert, J. (1999). *The teaching gap: Best ideas from the world's teachers for improving education in the classroom*. New York: The Free Press.

Sugiyama, Y. (2008). *Shotoka sugakukakyokiku josetsu-Sugiyama Yoshishige kogi hikki* (Elementary mathematics education beginning theory—lecture records of Yoshishige Sugiyama). Tokyo: Toyokan Shoseki.

Takahashi, A., & Yoshida, M. (2004). Ideas for establishing lesson study communities. *Teaching Children Mathematics, 10*(9), 436–443.

Takahashi, A., Watanabe, T., Yoshida, M., & Wang-Iverson, P. (2005). Improving content and pedagogical knowledge through kyozaikenkyu. In P. Wang-Iverson & M. Yoshida (Eds.), *Building our understanding of lesson study*. Philadelphia: Research for Better Schools.

Watanabe, T., Takahashi, A., & Yoshida, M. (2008). Kyozaikenkyu: A critical step for conducting effective lesson study and beyond. In F. Arbaugh & P. M. Taylor (Eds.), *AMTE monograph 5: Inquiry into mathematics teacher education*, pp. 131–142.

Yoshida, M. (1999). *Lesson Study: A case study of a Japanese approach to improving instruction through school-based teacher development.* Unpublished doctoral dissertation, The University of Chicago, Chicago.

Yoshida, M. (2008). Exploring ideas for a mathematics teacher educator's contribution to lesson study: Toward improving teachers' mathematical content and pedagogical knowledge. In D. Tirosh & T. Wood (Eds.), *The international handbook of mathematics teacher education: Vol. 2. Tools and processes in mathematics teacher education* (pp. 85–106). Rotterdam: Sense Publishers.

Final Thoughts

From the authors in this volume, it is apparent that lesson study is gaining interest around the globe; however, acceptance and implementation of the model as a viable and transformative professional learning tool in cultures outside of Japan will undoubtedly be a measured process. In this volume, we attempted to identify and present examples of research on lesson study and representative implementations that are occurring at various sites and in diverse environments.

Collectively the chapters begin to unpack the complexity of lesson study as a process and as a tool for teacher-centered, classroom-based professional growth. As Takahashi states, "The education leaders and researchers who support teachers through the lesson study process should keep in mind that while the *idea* of lesson study is simple…lesson study is not an easy jump for those teachers who have never experienced it." Through these varied lenses we begin to capture the intricacy of lesson study. The outcomes described in each chapter provide evidence both of recognized benefits and obstacles yet to be overcome.

Specific challenges addressed by the authors include the lack of a cadre of experienced lesson study teachers, teacher educators, and researchers (outside experts/knowledgeable others) that limit the depth, success, and expansion of lesson study in non-Japanese educational settings. The need for individuals with the necessary level of expertise is articulated clearly by Watanabe, Takahashi, and Yoshida and Jackson in this volume, but the dilemma of how to develop informed individuals who can successfully support lesson study groups is not resolved. A particularly challenging aspect of becoming an outside expert is acquiring a *learning stance* (Watanabe, this volume) where leaders and facilitators see themselves as colearners with the participating teachers. This change in *disposition* may not be easy for many experienced teacher educators who have been accustomed to directing professional learning as an exercise in imparting knowledge to teachers. In a similar vein, the learning disposition of participating teachers (Kamina and Tinto; Lewis, this volume) is integral to the success of a lesson study. Teachers who are mandated to participate without first having an opportunity to understand and develop some trust in the potential of this approach are less likely to benefit, and may instill a negative attitude that undercuts the experience for other members of their lesson study community. The buy-in that is evident in descriptions of the experience of *Japanese* teachers is not always present in other cultures.

L. C. Hart et al. (eds.), *Lesson Study Research and Practice in Mathematics Education*,
DOI 10.1007/978-90-481-9941-9,

A second major challenge is that of investigating instructional materials (*kyozai-kenkyu*), designing tasks (Doig et al. in this volume), and developing the research lesson (Takahashi). These are all complex components of the lesson study cycle that, if done superficially, often dilute the benefits and outcomes of lesson study. Many teachers have never engaged in multifaceted, in-depth professional work such as this, and without adequate support will not progress beyond the surface.

Larger systemic challenges include curricula that are unfocused and difficult to use within a lesson study environment; demands on teachers in some countries, especially in assessment and accountability; and priorities within school districts such as budget constraints, scheduling difficulties, and an inability to find sufficient amounts of time.

But the chapters in this book give us hope. In an effort to expose student thinking, lesson study concurrently supports the principles and philosophy of mathematical experiences in which *children* explore and dig deeply into concepts. It also supports development of teacher knowledge of content, pedagogy, and children's thinking. It appears from several chapters in the book that it is possible to re-create some form of lesson study in a novice environment and teachers can (and do) change and learn as a result of participating in these newly formed lesson study groups.

As the respondents to the chapters pointed out, by experiencing lesson study, teachers and others who participate in it become more experienced, thus become more knowledgeable and will be able to implement it better in the future. For example, Potari identifies the usefulness of lesson study for preservice teachers as they experience an integration of theory and practice, a learning goal often not acquired in professional learning experiences. Watanabe points out the usefulness of lesson study as a research tool for exposing and studying teacher learning as well as for teachers to research their own practice. As a professional learning model, lesson study *integrates* the typically discrete components of content knowledge, pedagogical knowledge, curricular knowledge, and knowledge of children's thinking. And, lesson study provides an avenue for teachers to develop and sustain an inquiry stance (Lewis; Sack and Vasquez, this volume).

Where should we go from here? Clearly, we need long-term, whole-school sites where we can study the impact on student learning. We also need to recognize that adjusting an existing system of professional learning undoubtedly will take time, and small shifts and adjustments may go unnoticed. Certainly, a future agenda item should be to examine the changes. Do we make changes in the system to accommodate the challenges listed above? Or, do we modify lesson study to work around the challenges? Or, does the whole system need to shift together?

Index

L. C. Hart et al. (eds.), *Lesson Study Research and Practice in Mathematics Education*, DOI 10.1007/978-90-481-9941-9, © Springer Science+Business Media B.V. 2011

T

V

W